Student Resource Manual
to accompany
Precalculus
with Calculus Previews

FOURTH EDITION

by Dennis G. Zill and Jacqueline M. Dewar

W. Scott Wright
Loyola Marymount University

Carol D. Wright

JONES AND BARTLETT PUBLISHERS
Sudbury, Massachusetts
BOSTON   TORONTO   LONDON   SINGAPORE

World Headquarters
Jones and Bartlett Publishers
40 Tall Pine Drive
Sudbury, MA 01776
978-443-5000
info@jbpub.com
www.jbpub.com

Jones and Bartlett Publishers Canada
6339 Ormindale Way
Mississauga, Ontario L5V 1J2
CANADA

Jones and Bartlett Publishers International
Barb House, Barb Mews
London W6 7PA
UK

Jones and Bartlett's books and products are available through most bookstores and online booksellers. To contact Jones and Bartlett Publishers directly, call 800-832-0034, fax 978-443-8000, or visit our website, www.jbpub.com.

Production Credits
Acquisition Editor: Timothy Anderson
Production Director: Amy Rose
Marketing Manager: Andrea DeFronzo
Manufacturing Buyer: Therese Connell
Editorial Assistant: Laura Pagluica
Composition: Northeast Compositors, Inc.
Cover Design: Anne Spencer
Cover Image: © Dynamic Graphics/Jupiterimages
Printing and Binding: Malloy, Inc.
Cover Printing: Malloy, Inc.

ISBN-13: 978-0-7637-4693-3
ISBN-10: 0-7637-4693-2

Printed in the United States of America
11 10 09 08 07   10 9 8 7 6 5 4 3 2 1

Contents

Preface

This manual is intended to accompany *Precalculus with Calculus Previews*, Fourth Edition, by Dennis G. Zill and Jacqueline M. Dewar. It consists of five parts, described below.

Topics in Algebra This part consists of short discussions of appropriate topics from a prerequisite algebra course (such as synthetic division), as well as topics intended to assist the student in becomming a more effective problem solver (such as implicit conditions in a word problem). See the Table of Contents for a complete list of topics covered.

Use of a Calculator This part is not intended to be a comprehensive manual on the use of a graphing calculator in a precalculus course. While much of the material discussed will be pertinent to any graphing calculator, the references in this manual will be to the TI-84 family of calculators. After a few brief comments on the use of the TI-84 calculator, this manual will focus on how to use the calculator to either assist in the solution of some of the problems in the text or to check that your solution is correct or at least reasonable.

Basic Skills This is a list for each section in the text of the skills needed to solve the more manipulative problems in the section.

Selected Solutions Here, complete solutions to every third problem (starting with Problem 3) in each section of the text are given. Solutions are not provided for **Calculator Problems** or for the problems **For Discussion**.

Final Examination Answers As indicated by the title of this part, the answers are provided for each of the 62 problems on the final examination, which is found on pages 380–383 in the text.

PART 1

Topics in Algebra

Topics in Algebra

1 | Multiplication by an Unknown in an Inequality

By (i) in the **Properties of Inequalities** in Section 1.1 of the text, there is no problem adding to (or subtracting from) both sides of an inequality an expression containing the variable. For example, in $3x > 2x - 4$, it is fine to subtract $2x$ from both sides and conclude that $3x - 2x > 2x - 4 - 2x$ or $x > -4$.

In an inequality like $x(x-1)/(x-2) \leq 0$ it is tempting to multiply both sides by $x - 2$ resulting in $x(x-1) \leq 0$. *This is incorrect* because, by (iii) in the **Properties of Inequalities** in Section 1.1 of the text, $x - 2$ will be negative for some values of x, which will result in the inequality changing from $\leq$ to $\geq$. *Thus, you should never multiply both sides of an inequality by an expression containing a variable that could cause the expression to be negative.* On the other hand, if you are confident that an expression is always positive, there is no problem multiplying or dividing by that expression in an inequality. For example, $(2x + 1)/(x^2 + 4) > 0$ is equivalent to $2x + 1 > 0$, because both sides of the inequality can be multiplied by $x^2 + 4$, which is positive for all values of x.

2 | Division by Zero When Solving an Equation

Of course, you know that you should never divide by zero. Sometimes, however, you may unwittingly do so. For example, if you attempt to solve the equation

$$(x - 3)^2 = 4(x - 3)$$

by dividing both sides by $x - 3$, you may have divided by zero. In fact, in this case you have, since the resulting equation, $x - 3 = 4$, has only the solution $x = 7$. This is a solution

3

of the original equation, but another solution, $x = 3$, has been lost. This is because when $x = 3$, $x - 3 = 0$, and you have divided by $x - 3$. A way to solve the equation without dividing by zero involves factoring and is shown here:

$$(x - 3)^2 = 4(x - 3)$$
$$(x - 3)^2 - 4(x - 3) = 0$$
$$(x - 3)(x - 3 - 4) = 0$$
$$(x - 3)(x - 7) = 0.$$

Thus, $x = 3$ and $x = 7$ are solutions of the original equation.

3   Implicit Conditions in a Word Problem

Frequently, a word problem will contain a condition or conditions that are not explicitly stated as part of the problem, but are implicit in the nature of the problem. For example, if t, representing time, is the unknown in a problem, then it will usually be the case that an implicit condition in the problem is $t \geq 0$. This means that if, say, $t = -4$ arises in the solution of an equation or inequality, it can and should be ignored and only nonnegative values of t should be considered.

As another, more detailed example, consider the following geometric problem:

Suppose the area of the ring between a pair of concentric circles is to be at most $9\pi\,\text{cm}^2$. Find the possible dimensions for the radius of the inner circle if the radius of the outer circle is fixed at 5 cm. [See the figure below.]

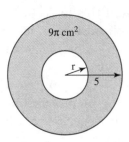

The solution of this problem involves the inequalities

$$\pi 25 - \pi r^2 < 9\pi$$
$$16 - r^2 < 0$$
$$(4 + r)(4 - r) < 0,$$

which leads to a sign chart involving $r = -4$ and $r = 4$. Although it is not explicitly stated, $r > 0$ is an implicit condition in this problem. Thus, we can ignore $r = -4$ and consider only positive values of r.

$r + 4$	$+$	$+$	$+$
$r - 4$	$-$	0	$+$
$(r + 4)(r - 4)$	$-$	0	$+$

From the graph we conclude that r must be greater than 4 for the area of the ring to be at most 9 cm^2. But, wait a minute! If we read the problem more carefully, we see that another implicit condition is $r < 5$, so we should have stopped the previous line diagram at $r = 5$, as shown here.

The correct solution of the problem is $4 < r < 5$.

4   Extraneous Roots

When solving an algebraic equation, a number that appears to be a solution may, in fact, not satisfy the original equation and hence not be a solution. We call such a number an *extraneous solution* or *extraneous root*. We will discuss two instances that can give rise to extraneous solutions. The first instance occurs when both sides of an equation are multiplied by an expression that is 0 for some value or values of the unknown. In this case, it is *essential* to check any solutions that have been obtained—they may or may not be solutions of the original equation. For example, multiplying both sides of the equation

$$2 - \frac{1}{x+1} = \frac{x}{x+1}$$

by $x + 1$ yields

$$(x+1)\left(2 - \frac{1}{x+1}\right) = (x+1)\left(\frac{x}{x+1}\right)$$

$$2x + 2 - 1 = x.$$

The solution of the last equation is $x = -1$, but this is not a solution of the original equation because substitution into the original equation results in division by 0, which is not permissible.

   Note: Extraneous solutions will not be obtained when both sides of an equation are multiplied by an expression that can never be 0. For example, when solving

$$\frac{x}{x^2+1} = \frac{1}{2},$$

both sides can be multiplied by $2(x^2 + 1)$ resulting in $2x = x^2 + 1$, $(x - 1)^2 = 0$, and $x = 1$. The number 1 cannot be an extraneous solution because both sides of the original equation were multiplied by the expression $2(x^2 + 1)$, which can never be 0. This note is not intended to discourage you from checking your answer. *This should always be done.*

   The second instance that can result in extraneous solutions occurs when both sides of an equation are squared. For example, squaring both sides of the equation

$$\sqrt{3x + 4} = x - 2,$$

results in

$$3x + 4 = (x - 2)^2$$
$$3x + 4 = x^2 - 4x + 4$$
$$x^2 - 7x = 0$$
$$x(x - 7) = 0.$$

Thus, it appears that $x = 7$ and $x = 0$ are solutions of the equation. In fact, substitution of $x = 7$ into the original equation yields the true statement $5 = 5$. However, substitution of $x = 0$ into the original equation yields $2 = -2$, which is false. In this case, $x = 0$ is an extraneous solution and the only solution of the original equation is $x = 7$.

5   Factorial Notation

Factorial notation is used to represent a descending product of nonnegative integers. The symbol $r!$ is defined for any nonnegative integer r by

$$1! = 1,$$
$$2! = 2 \cdot 1 = 2,$$
$$3! = 3 \cdot 2 \cdot 1 = 6,$$
$$4! = 4 \cdot 3 \cdot 2 \cdot 1 = 24,$$

and so on. In general

$$n! = n(n - 1)(n - 2) \cdots 3 \cdot 2 \cdot 1,$$

where n is a positive integer. There are a number of reasons that justify the additional definition of

$$0! = 1.$$

We will see one of these reasons in the section on the binomial theorem.

■ **EXAMPLE**   Simplify $7!/4!$.

Solution   Using the definition of factorial we have

$$\frac{7!}{4!} = \frac{7 \cdot 6 \cdot 5 \cdot 4!}{4!} = 7 \cdot 6 \cdot 5 = 210.$$

□

■ **EXAMPLE**   Simplify

$$\frac{n!(n + 1)}{(n - 1)!}.$$

Solution   Using the definition of factorial we can write the numerator as

$$n!(n + 1) = (n + 1)n! = (n + 1)n(n - 1) \cdots 3 \cdot 2 \cdot 1 = (n + 1)n(n - 1)!.$$

Then,

$$\frac{n!(n + 1)}{(n - 1)!} = \frac{(n + 1)n(n - 1)!}{(n - 1)!} = (n + 1)n.$$

□

6   Pascal's Triangle

Pascal's triangle is a triangular array of positive integers. We will see an application of Pascal's triangle in the next section on the Binomial Theorem. The first five rows of the triangle are shown here.

$$
\begin{array}{ccccccccc}
 & & & & 1 & & & & \\
 & & & 1 & & 1 & & & \\
 & & 1 & & 2 & & 1 & & \\
 & 1 & & 3 & & 3 & & 1 & \\
1 & & 4 & & 6 & & 4 & & 1 \\
\end{array}
$$

Observe that the outer sides of the triangle consist entirely of 1s, and that each number in the interior of the triangular array is the sum of the two numbers above to the left and to the right of that number. For example the first 4 in the fifth row is the sum of 1 and 3, the two numbers above the four.

◾ **EXAMPLE**   Find the next row in Pascal's triangle.

Solution   The diagram below shows how to obtain the sixth row of Pascal's triangle by using the fifth row.

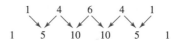

□

7   Binomial Theorem

When $(a + b)^n$ is expanded, or multiplied out, for an arbitrary positive integer n, the exponents of a and b follow a definite pattern. For example, from

$$
\begin{aligned}
(a + b)^2 &= a^2 + 2ab + b^2, \\
(a + b)^3 &= a^3 + 3a^2b + 3ab^2 + b^3, \\
(a + b)^4 &= a^4 + 4a^3b + 6a^2b^2 + 4ab^3 + b^4,
\end{aligned}
$$

we see that the exponents of a decrease by 1, starting with the first term, whereas the exponents of b increase by 1, starting with the second term. To extend this pattern, we consider the first and last terms to be multiplied by b^0 and a^0, respectively; that is,

$$
(a + b)^4 = a^4b^0 + 4a^3b + 6a^2b^2 + 4ab^3 + a^0b^4.
$$

We also note that the sum of the exponents in each term of the expansion of $(a + b)^4$ is 4.

The following two examples illustrate the use of the product formulas given at the beginning of this section.

■ **EXAMPLE**   Expand $(x^2 + 2k)^2$.

Solution Identifying $a = x^2$ and $b = 2k$, it follows from the formula for $(a + b)^2$ at the start of this section that

$$(x^2 + 2k)^2 = (x^2)^2 + 2(x^2)(2k) + (2k)^2 = x^4 + 4x^2 k + 4k^2.$$

$\square$

■ **EXAMPLE**   Expand $(\sqrt{x} - 4y^2)^3$.

Solution Identifying $a = \sqrt{x}$ and $b = -4y^2$, it follows from the formula for $(a + b)^3$ at the start of this section that

$$\begin{aligned}
(\sqrt{x} - 4y^2)^3 &= (\sqrt{x})^3 + 3(\sqrt{x})^2(-4y^2) + 3(\sqrt{x})(-4y^2)^2 + (-4y^2)^3 \\
&= x^{3/2} - 12xy^2 + 48x^{1/2}y^4 - 64y^6.
\end{aligned}$$

$\square$

Note: In the last example we expressed $(\sqrt{x})^3$ in the form $x^{3/2}$ and $\sqrt{x}$ in the form $x^{1/2}$. That is, we switched from the use of radical notation to the use of fractional exponents. This is generally good practice, especially in calculus, because fractional exponents are usually easier to manipulate than expressions containing radicals.

As discussed previously, the exponents in a binomial expansion follow a pattern. The same is true of the coefficients. In fact, in the case of the coefficients, this pattern is given by Pascal's triangle, discussed in the preceding section of this manual. For example,

$$(a + b)^5 = a^5 + 5a^4 b + 10a^3 b^2 + 10a^2 b^3 + 5ab^4 + b^5,$$

and the coefficients constitute the sixth row of Pascal's triangle.

■ **EXAMPLE**   Expand $(x - y)^7$.

Solution We begin by expanding Pascal's triangle through the eighth row:

$$
\begin{array}{ccccccccccccccc}
 & & & & & & & 1 & & & & & & & \\
 & & & & & & 1 & & 1 & & & & & & \\
 & & & & & 1 & & 2 & & 1 & & & & & \\
 & & & & 1 & & 3 & & 3 & & 1 & & & & \\
 & & & 1 & & 4 & & 6 & & 4 & & 1 & & & \\
 & & 1 & & 5 & & 10 & & 10 & & 5 & & 1 & & \\
 & 1 & & 6 & & 15 & & 20 & & 15 & & 6 & & 1 & \\
1 & & 7 & & 21 & & 35 & & 35 & & 21 & & 7 & & 1
\end{array}
$$

Now, identifying a with x and b with $-y$, it follows that

$$\begin{aligned}
(x - y)^7 &= x^7 + 7x^6(-y) + 21x^5(-y)^2 + 35x^4(-y)^3 + 35x^3(-y)^4 + 21x^2(-y)^5 + 7x(-y)^6 + (-y)^7 \\
&= x^7 - 7x^6 y + 21x^5 y^2 - 35x^4 y^3 + 35x^3 y^4 - 21x^2 y^5 + 7xy^6 - y^7.
\end{aligned}$$

$\square$

While Pascal's triangle is easy to use when expanding a binomial expression of the form $(a+b)^n$ for relatively small values of n, it would be very time consuming to generate enough rows of Pascal's triangle to expand $(a + b)^{14}$. In this case it is more convenient to use the general formula for the expansion of $(a + b)^n$ given by the **binomial theorem**.

BINOMIAL THEOREM For any positive integer n,

$$(a + b)^n = a^n + \frac{n}{1!} a^{n-1}b + \frac{n(n-1)}{2!} a^{n-2}b^2$$
$$+ \cdots + \frac{n(n-1)\cdots(n-k+1)}{k!} a^{n-k}b^k + \cdots + b^n.$$

∎

We can write the binomial theorem in a more compact form using summation notation (discussed in Section 3.7 in the text) and without the ellipses ($\cdots$) by using the fact that for any integer $0 \le k \le n$,

$$n(n-1)\cdots(n-k-1) = \frac{n(n-1)\cdots(n-k+1)(n-k)(n-k-1)\cdots 3 \cdot 2 \cdot 1}{(n-k)(n-k-1)\cdots 3 \cdot 2 \cdot 1}$$
$$= \frac{n!}{(n-k)!}.$$

Using this fact, we then have

$$(a + b)^n = \frac{n!}{0!(n-0)!} a^{n-0}b^0 + \frac{n!}{1!(n-1)!} a^{n-1}b^1$$
$$+ \cdots + \frac{n!}{k!(n-k)!} a^{n-k}b^k + \cdots + \frac{n!}{n!(n-n)!} a^{n-n}b^{n-0}$$
$$= \sum_{k=0}^{n} \frac{n!}{k!(n-k)!} a^{n-k}b^k.$$

Note that at this point we require that $0! = 1$.

The binomial theorem now provides an alternative to Pascal's triangle for an expression of the form $(a + b)^n$.

EXAMPLE Use the binomial theorem to expand $(a + b)^4$.

Solution From the binomial theorem we have

$$(a + b)^4 = a^4 + \frac{4}{1!} a^{4-1}b + \frac{4(3)}{2!} a^{4-2}b^2 + \frac{4(3)(2)}{3!} a^{4-3}b^3 + \frac{4(3)(2)(1)}{4!} b^4$$
$$= a^4 + 4a^3b + \frac{12}{2} a^2b^2 + \frac{24}{6} ab^3 + \frac{24}{24} b^4$$
$$= a^4 + 4a^3b + 6a^2b^2 + 4ab^3 + b^4.$$

□

EXAMPLE Find the sixth term of the expansion of $(x^2 - 2y)^7$.

Solution Letting $n = 7$ in the summation notation form of the binomial theorem we have

$$(a + b)^7 = \sum_{k=0}^{7} \frac{7!}{k!(7-k)!} a^{7-k}b^k.$$

Since the sixth term corresponds to $k = 5$ (because k starts with 0 instead of 1), the sixth term is

$$\frac{7!}{5!(7-5)!} a^{7-5}b^5 = \frac{7!}{5!(2!)} a^2b^5 = \frac{7(6)}{2(1)} a^2b^5 = 21a^2b^5.$$

Now, identifying $a = x^2$ and $b = -2y$, we see that the sixth term is

$$21(x^2)^2(-2y)^5 = 21x^4(-32y^5) = -672x^4y^5.$$

□

8   Factoring

Factoring polynomials is the reverse of multiplying polynomials. That is, given a polynomial like $5x^3 + 6x^2 - 29x - 6$, we would like to be able to express it as a product of simpler polynomials. For example,

$$5x^3 + 6x^2 - 29x - 6 = (5x + 1)(x - 2)(x + 3).$$

Factoring can be very useful when solving equations or evaluating limits.

8.1   Common Factors and Grouping

In general, the first step in factoring any algebraic expression is to determine whether the terms have a common factor.

EXAMPLE   Factor $6x^4y^4 - 4x^2y^2 + 10xy^3 - 2xy^2$.

Solution Since $2xy^2$ is a common factor of the terms, we have

$$6x^4y^4 - 4x^2y^2 + 10xy^3 - 2xy^2 = 2xy^2(3x^3y^2) - 2xy^2(2x) + 2xy^2(5y) - 2xy^2$$
$$= 2xy^2(3x^3y^2 - 2x + 5y - 1).$$

$\square$

When the terms of an expression do not have a common factor, it may still be possible to factor by *grouping* the terms in an appropriate manner.

EXAMPLE   Factor $x^2 + 2xy - x - 2y$.

Solution Grouping the first two terms and the last two terms gives

$$x^2 + 2xy - x - 2y = (x^2 + 2xy) + (-x - 2y) = x(x + 2y) + (-1)(x + 2y).$$

We observe the common factor $x + 2y$ and complete the factorization as

$$x^2 + 2xy - x - 2y = (x - 1)(x + 2y).$$

$\square$

8.2   Use of Factorization Formulas

By reversing several product formulas we obtain the following important factorization formulas.

Factorization Formulas	
(i) **Square of a Sum:**	$X^2 + 2XY + Y^2 = (X + Y)^2$
(ii) **Square of a Difference:**	$X^2 - 2XY + Y^2 = (X - Y)^2$
(iii) **Difference of Two Squares:**	$X^2 - Y^2 = (X - Y)(X + Y)$
(iv) **Sum of Two Cubes:**	$X^3 + Y^3 = (X + Y)(X^2 - XY + Y^2)$
(v) **Difference of Two Cubes:**	$X^3 - Y^3 = (X - Y)(X^2 + XY + Y^2)$

We have used capital letters in these formulas to clarify our work when we apply the formulas.

◼ **EXAMPLE**   Factor $16x^4y^2 - 25$.

Solution   This is the difference of two squares. Thus, from **(iii)**, with $X = 4x^2y$ and $Y = 5$, we have

$$16x^4y^2 - 25 = (4x^2y)^2 - (5)^2 = (4x^2y - 5)(4x^2y + 5).$$

◻

◼ **EXAMPLE**   Factor $8a^3 + 27b^6$.

Solution   This is the sum of two cubes. Thus, from **(iv)**, with $X = 2a$ and $Y = 3b^2$, we have

$$8a^3 + 27b^6 = (2a)^3 + (3b^2)^3 = (2a + 3b^2)[(2a)^2 - (2a)(3b^2) + (3b^2)^2]$$
$$= (2a + 3b^2)(4a^2 - 6ab^2 + 9b^4).$$

◻

Observe in the above factorization formulas that there is no formula for the sum of two squares. An expression of the form $X^2 + Y^2$ never factors in the real number system. On the other hand, formulas **(iii)**–**(v)** indicate that the difference of two squares and the sum and difference of two cubes always factor, provided that we do not restrict the coefficients to the set of integers. For example, using **(iii)** to factor $x^2 - 5$, we identify $X = x$ and $Y = \sqrt{5}$ so that

$$x^2 - 5 = x^2 - (\sqrt{5})^2 = (x - \sqrt{5})(x + \sqrt{5}).$$

For the remainder of this section, however, we will seek only polynomial factors with *integer* solutions.

8.3   Factoring Quadratic Polynomials

It is sometimes possible to factor the quadratic polynomial $ax^2 + bx + c$, where a, b, and c are integers, as

$$(Ax + B)(Cx + D),$$

where A, B, C, and D are also integers. Initially, to simplify our discussion we assume that the quadratic polynomial has as its leading coefficient $a = 1$. If $x^2 + bx + c$ has a factorization using integer coefficients, then it will be of the form

$$(x + B)(x + D),$$

where B and D are integers. Finding the product and comparing coefficients,

$$(x + B)(x + D) = x^2 + \overbrace{(B + D)}^{B+D=b}x + \underbrace{BD = x^2 + b\,x + c}_{BD=c},$$

we see that

$$B + D = b \qquad \text{and} \qquad BD = c.$$

Thus, to factor $x^2 + bx + c$ with integer coefficients, we list all possible factorizations of c as a product of two integers B and D. We then check which, if any, of the sums $B + D$ equals b.

■ **EXAMPLE** Factor $x^2 - 9x + 18$.

Solution With $b = -9$ and $c = 18$, we look for integers B and D such that

$$B + D = -9 \quad \text{and} \quad BD = 18.$$

We can write 18 as a product BD in the following ways:

$$1(18), \quad 2(9), \quad 3(6), \quad (-1)(-18), \quad (-2)(-9), \quad \text{or} \quad (-3)(-6),$$

Since -9 is the sum of -3 and -6, the factorization is

$$x^2 - 9x + 18 = (x - 3)(x - 6).$$

□

Note: Once you have obtained a factorization, you should *always* check your work by multiplying out the factors.

As we see in the next example, it is always possible that a quadratic expression will not factor.

■ **EXAMPLE** Factor $x^2 + 3x - 2$.

Solution The number -2 can be written as a product of integers in two ways: $(-1)(2)$ and $(1)(-2)$. Since neither $-1 + 2$ nor $1 + (-2)$ is equal to 3, the expression $x^2 + 3x - 2$ cannot be factored using integer coefficients.

□

It is more complicated to factor the general quadratic polynomial $ax^2 + bx + c$, when $a \neq 1$, since, in this case, we must consider factors of a as well as of c. Finding the product and comparing the coefficients in

$$(Ax + B)(Cx + D) = \underbrace{AC}_{AC=a}x^2 + (\underbrace{AD + BC}_{AD+BC=b})x + \underbrace{BD}_{BD=c} = ax^2 + bx + c,$$

we see that $ax^2 + bx + c$ factors as $(Ax + B)(Cx + D)$ if

$$AC = a, \quad AD + BC = b, \quad \text{and} \quad BD = c.$$

■ **EXAMPLE** Factor $2x^2 + 11x - 6$.

Solution The factors will be

$$(2x + \underline{\quad})(1x + \underline{\quad}),$$

where the blanks are to be filled with a pair of integers B and D whose product BD equals -6. Possible pairs are

$$
\begin{array}{llll}
1 \text{ and } -6, & -1 \text{ and } 6, & 3 \text{ and } -2, & -3 \text{ and } 2, \\
-6 \text{ and } 1, & 6 \text{ and } -1 & -2 \text{ and } 3, & 2 \text{ and } -3.
\end{array}
$$

Now we must check to see if one of the pairs gives 11 as the value of $AD + BC$ (the coefficient of the middle term), where $A = 2$ and $C = 1$. We find

$$2(6) + (-1)(1) = 11,$$

so $2x^2 + 11x - 6 = (2x - 1)(x + 6)$.

□

This general method can be applied to expressions of the form $ax^2 + bxy + cy^2$, where a, b, and c are integers.

■ **EXAMPLE**   Factor $15x^2 + 17xy + 4y^2$.

Solution The possible factors have the form

$$(5x + \underline{}y)(3x + \underline{}y) \quad \text{or} \quad (15x + \underline{}y)(1x + \underline{}y).$$

There is no need to consider the cases

$$(-5x + \underline{}y)(-3x + \underline{}y) \quad \text{or} \quad (-15x + \underline{}y)(-x + \underline{}y)$$

because we can factor -1 out from each term and $(-1)(-1) = 1$, so these two cases are equivalent to the first two cases. Now, the blanks must be filled with a pair of integers whose product is 4. Possible pairs are

$$1 \text{ and } 4, \quad -1 \text{ and } -4, \quad 2 \text{ and } 2, \quad -2 \text{ and } -2.$$

We check pairs of numbers to see which combination, if any, gives a coefficient of 17 for the middle term. We find

$$15x^2 + 17xy + 4y^2 = (5x + 4y)(3x + y).$$

□

■ **EXAMPLE**   Factor $2x^4 + 11x^2 + 12$.

Solution Letting $X = x^2$, we can regard this expression as a quadratic polynomial in the variable X,

$$2X^2 + 11X + 12.$$

We then factor this quadratic polynomial. Since all of the coefficients of $2X^2 + 11X + 12$ are positive, we can assume that the factors will have the form

$$(X + \underline{})(2X + \underline{}),$$

where the blanks are to be filled with a pair of positive integers whose product is 12. The possible pairs are

$$1 \text{ and } 12, \quad 2 \text{ and } 6, \quad 3 \text{ and } 4, \quad 12 \text{ and } 1, \quad 6 \text{ and } 2, \quad 4 \text{ and } 3.$$

We check each pair to see which combination, if any, gives a coefficient of 11 for the middle term. We find

$$2X^2 + 11X + 12 = (X + 4)(2X + 3).$$

Substituting x^2 for X gives

$$2x^4 + 11x^2 + 12 = (x^2 + 4)(2x^2 + 3).$$

□

In the preceding example you should consider the possibility that $x^2 + 4$ or $2x^2 + 3$ factor. In this case, since each is a *sum* of squares, neither factors.

9   Rationalize the Numerator

When we remove radicals from the numerator or denominator of a fraction, we say that we are **rationalizing**. In algebra, we usually rationalize the denominator, but in calculus it is sometimes important to rationalize the numerator. The procedure of rationalizing involves multiplying the fraction by 1, written in a special way. For example, to rationalize $2/\sqrt{3}$ we multiply by $\sqrt{3}/\sqrt{3}$:

$$\frac{2}{\sqrt{3}} = \frac{2}{\sqrt{3}}\left(\frac{\sqrt{3}}{\sqrt{3}}\right) = \frac{2\sqrt{3}}{3}.$$

If a fraction contains an expression such as $2 + \sqrt{x}$, we make use of the fact that the product of $2 + \sqrt{x}$ and its *conjugate* $2 - \sqrt{x}$ contains no radicals:

$$\frac{x}{2+\sqrt{x}} = \frac{x}{2+\sqrt{x}}\left(\frac{2-\sqrt{x}}{2-\sqrt{x}}\right) = \frac{2x - x\sqrt{x}}{4-x}.$$

We proceed in a similar fashion to rationalize the numerator of a fraction containing radicals:

$$\frac{\sqrt{x-2}-\sqrt{y}}{x+y} = \frac{\sqrt{x-2}-\sqrt{y}}{x+y}\left(\frac{\sqrt{x-2}+\sqrt{y}}{\sqrt{x-2}+\sqrt{y}}\right) = \frac{x-2-y}{(x+y)(\sqrt{x-2}+\sqrt{y})}.$$

10   Rational Expressions

A quotient of two polynomials is called a **rational expression**. For example,

$$\frac{2x^2 + 5}{x+1} \qquad \text{and} \qquad \frac{3}{2x^3 - x + 8}$$

are rational expressions. In solving problems, we often must combine rational expressions and then simplify the results. To do this we use the following properties of numbers, which are valid provided each denominator is nonzero.

Properties of Fractions

(i)   **Cancellation:**
$$\frac{ac}{bc} = \frac{a}{b}$$

(ii)   **Addition or Subtraction:**
$$\frac{a}{b} \pm \frac{c}{b} = \frac{a \pm c}{b}$$

(iii)   **Multiplication:**
$$\frac{a}{b} \cdot \frac{c}{d} = \frac{ac}{bd}$$

(iv)   **Division:**
$$\frac{a}{b} \div \frac{c}{d} = \frac{a}{b} \cdot \frac{d}{c}$$

■ **EXAMPLE**   Simplify

$$\frac{2x^2 - x - 1}{x^2 - 1}.$$

Solution   We factor the numerator and denominator and cancel common factors (noted in gray type) using the cancellation property **(i)**:

$$\frac{2x^2 - x - 1}{x^2 - 1} = \frac{(2x + 1)(x - 1)}{(x + 1)(x - 1)} = \frac{2x + 1}{x + 1}.$$

$\square$

Note that in the previous example the cancellation of the common factor $x - 1$ is valid only for those values of x such that $x - 1$ is nonzero; that is, for $x \neq 1$. However, since the expression $(2x^2 - x - 1)/(x^2 - 1)$ is not defined for $x = 1$, our simplification is valid for all real numbers in the domain of the variable x in the *original* expression. We emphasize that the equation

$$\frac{2x^2 - x - 1}{x^2 - 1} = \frac{2x + 1}{x + 1}$$

is not valid for $x = 1$, even though the right-hand side, $(2x + 1)/(x + 1)$ is defined for $x = 1$. Considerations of this sort are important when solving equations involving rational expressions.

Henceforth, we will assume without further comment that variables are restricted to values for which all denominators in an equation are nonzero.

EXAMPLE   Simplify

$$\frac{4x^2 + 11x - 3}{2 - 5x - 12x^2}.$$

Solution   We factor the numerator and denominator and cancel common factors (noted in gray type) using the cancellation property **(i)**:

$$\begin{aligned}
\frac{4x^2 + 11x - 3}{2 - 5x - 12x^2} &= \frac{(4x - 1)(x + 3)}{(1 - 4x)(2 + 3x)} \\
&= \frac{(4x - 1)(x + 3)}{-(4x - 1)(2 + 3x)} \\
&= -\frac{x + 3}{2 + 3x}.
\end{aligned}$$

$\square$

11   [Least Common Denominator]

In order to add or subtract rational expressions, we proceed just as we do when adding or subtracting fractions. We first find a common denominator and then apply **(ii)** in the **Properties of Fractions** on the previous page. Although any common denominator will do, less work is involved if we use the *least common denominator* (LCD). This is found by factoring each denominator completely and forming a product of the distinct factors, using each factor with the highest exponent with which it occurs in any single denominator.

EXAMPLE   Find the LCD of

$$\frac{1}{x^4 - x^2}, \qquad \frac{x + 2}{x^2 + 2x + 1}, \qquad \text{and} \qquad \frac{1}{x}.$$

Solution   Factoring the denominators in the rational expressions, we obtain

$$\frac{1}{x^2(x - 1)(x + 1)}, \qquad \frac{x + 2}{(x + 1)^2}, \qquad \text{and} \qquad \frac{1}{x}.$$

The distinct factors of the denominators are x, $x - 1$, and $x + 1$. We use each factor with the highest exponent with which it occurs in any single denominator. Thus, the LCD is

$$x^2(x - 1)(x + 1)^2.$$

■ **EXAMPLE**   Combine and simplify

$$\frac{x}{x^2 - 4} + \frac{1}{x^2 + 4x + 4}.$$

Solution In factored form, the denominators are $(x - 2)(x + 2)$ and $(x + 2)^2$. Thus, the LCD is $(x - 2)(x + 2)^2$. We use **(i)** in the **Properties of Fractions** in reverse to rewrite each rational expression with the LCD as a common denominator:

$$\frac{x}{x^2 - 4} = \frac{x}{(x - 2)(x + 2)} = \frac{x(x + 2)}{(x - 2)(x + 2)(x + 2)}$$

$$\frac{1}{x^2 + 4x + 4} = \frac{1}{(x + 2)^2} = \frac{1(x - 2)}{(x + 2)^2(x - 2)}.$$

Then, using **(ii)**, we add and simplify:

$$\frac{x}{x^2 - 4} + \frac{1}{x^2 + 4x + 4} = \frac{x(x + 2)}{(x - 2)(x + 2)^2} + \frac{x - 2}{(x - 2)(x + 2)^2}$$

$$= \frac{x(x + 2) + x - 2}{(x - 2)(x + 2)^2}$$

$$= \frac{x^2 + 2x + x - 2}{(x - 2)(x + 2)^2}$$

$$= \frac{x^2 + 3x - 2}{(x - 2)(x + 2)^2}.$$

□

To multiply or divide rational expressions, we apply **(iii)** or **(iv)** and then simplify.

■ **EXAMPLE**   Combine and simplify

$$\frac{x}{5x^2 + 21x + 4} \cdot \frac{25x^2 + 10x + 1}{3x^2 + x}.$$

Solution We factor, cancel, and then use **(iii)** in the **Properties of Fractions** to combine:

$$\frac{x}{5x^2 + 21x + 4} \cdot \frac{25x^2 + 10x + 1}{3x^2 + x} = \frac{x(25x^2 + 10x + 1)}{(5x^2 + 21x + 4)(3x^2 + x)}$$

$$= \frac{x(5x + 1)(5x + 1)}{(5x + 1)(x + 4)x(3x + 1)}$$

$$= \frac{5x + 1}{(x + 4)(3x + 1)}.$$

□

12 | Complex Fractions

A complex fraction is a fractional expression in which the numerator or denominator, or both, is itself a fraction. The examples below illustrate two techniques that can be used to simplify complex fractions.

■ **EXAMPLE**   Simplify

$$\frac{\dfrac{1}{x} - \dfrac{x}{x+1}}{1 + \dfrac{1}{x}}.$$

Solution [**Method I**] First we obtain a single rational expression for the numerator and the denominator:

$$\frac{1}{x} - \frac{x}{x+1} = \frac{1(x+1)}{x(x+1)} - \frac{x \cdot x}{x(x+1)} = \frac{x+1-x^2}{x(x+1)} = \frac{-x^2+x+1}{x(x+1)}$$

and

$$1 + \frac{1}{x} = \frac{x}{x} + \frac{1}{x} = \frac{x+1}{x}.$$

Thus,

$$\frac{\dfrac{1}{x} - \dfrac{x}{x+1}}{1 + \dfrac{1}{x}} = \frac{\dfrac{-x^2+x+1}{x(x+1)}}{\dfrac{x+1}{x}}.$$

Now we apply (**iv**), in the **Properties of Fractions**, to this quotient to obtain

$$\frac{\dfrac{-x^2+x+1}{x(x+1)}}{\dfrac{x+1}{x}} = \frac{-x^2+x+1}{x(x+1)} \cdot \frac{x}{x+1} = \frac{-x^2+x+1}{(x+1)^2}.$$

$\square$

An alternative method of simplifying a complex fraction assumes that the original fraction has no complex fractions in either its numerator or denominator. That is, the method described below would not apply to the complex fraction

$$\frac{\dfrac{x}{x-1} - \dfrac{\dfrac{1}{x} + x}{x+1}}{\dfrac{1}{x} + \dfrac{x}{x-1}}$$

because the numerator itself contains the complex fraction

$$\frac{\dfrac{1}{x} + x}{x+1}.$$

To describe an alternative method for simplifying complex fractions we introduce the notion of the *least common multiple* of polynomials. This is closely related to the notion of LCD. The least common multiple of a set of polynomials is obtained by factoring each polynomial completely and forming a product of the distinct factors, using each factor with the highest exponent with which it occurs in any factorization. We note that the LCD of a set of rational expressions is the least common multiple of the denominators of the expressions.

■ **EXAMPLE** Simplify

$$\frac{\dfrac{1}{x} - \dfrac{x}{x+1}}{1 + \dfrac{1}{x}}.$$

Solution [**Method II**] The denominators in the numerator of the original fraction are x and $x + 1$, and the denominator in the denominator of the original fraction is simply x. The least common multiple of these so called *secondary denominators* is $x(x + 1)$. We now multiply the original fraction by 1 written in the form $x(x+1)/x(x+1)$ and then simplify the result:

$$\frac{\dfrac{1}{x} - \dfrac{x}{x+1}}{1 + \dfrac{1}{x}} = \frac{\dfrac{1}{x} - \dfrac{x}{x+1}}{1 + \dfrac{1}{x}} \frac{x(x+1)}{x(x+1)} = \frac{(x+1) - x^2}{x(x+1) + (x+1)}$$

$$= \frac{-x^2 + x + 1}{(x+1)(x+1)} = \frac{-x^2 + x + 1}{(x+1)^2}$$

□

The techniques discussed above can often be applied to expressions containing negative exponents, as we see in the following example.

■ **EXAMPLE** Simplify $(a^{-1} + b^{-1})^{-1}$.

Solution [**Method I**] We first replace all negative exponents by the equivalent quotients and then use **Properties of Fractions** to simplify the resulting algebraic expression:

$$(a^{-1} + b^{-1})^{-1} = \frac{1}{a^{-1} + b^{-1}}$$

$$= \frac{1}{\dfrac{1}{a} + \dfrac{1}{b}} = \frac{1}{\dfrac{b+a}{ab}}$$

$$= \frac{ab}{b+a}.$$

Solution [**Method II**] We replace all negative exponents by the equivalent quotients and then multiply by ab/ab, the least common multiple of the secondary denominators:

$$(a^{-1} + b^{-1})^{-1} = \frac{1}{a^{-1} + b^{-1}}$$

$$= \frac{1}{\dfrac{1}{a} + \dfrac{1}{b}} \frac{ab}{ab}$$

$$= \frac{ab}{b+a}.$$

□

Our final example in this section illustrates a common rational expression occurring in calculus.

■ **EXAMPLE**   Simplify

$$\frac{\dfrac{1}{x+h} - \dfrac{1}{x}}{h}.$$

Solution [**Method I**] First we obtain a single rational expression for the numerator and then simplify the resulting fraction:

$$\frac{\dfrac{1}{x+h} - \dfrac{1}{x}}{h} = \frac{\dfrac{x-(x+h)}{(x+h)x}}{h} = \frac{\dfrac{-h}{x(x+h)}}{h}$$

$$= \frac{-h}{x(x+h)h} = \frac{-1}{x(x+h)}.$$

Solution [**Method II**] The least common multiple of the secondary denominators, $x+h$ and x, is $x(x+h)$. We multiply the numerator and denominator by this least common multiple and simplify:

$$\frac{\dfrac{1}{x+h} - \dfrac{1}{x}}{h} = \frac{\dfrac{1}{x+h} - \dfrac{1}{x}}{h} \cdot \frac{x(x+h)}{x(x+h)}$$

$$= \frac{x-(x+h)}{hx(x+h)}$$

$$= \frac{-h}{hx(x+h)}$$

$$= \frac{-1}{x(x+h)}.$$

□

13 Even and Odd Functions

As discussed in the text, a function f is even if it is symmetric with respect to the y-axis, and is odd if it is symmetric with respect to the origin. Using the tests for symmetry given in Section 2.2 of the text, we have the following:

> ### Tests for Even and Odd Functions
>
> **(i) Even Function:**
>   A function $f(x)$ is even if $f(-x) = f(x)$ for every x in the domain of f.
>
> **(ii) Odd Function:**
>   A function $f(x)$ is odd if $f(-x) = -f(x)$ for every x in the domain of f.

It should be easy to convince yourself graphically that the only function f that is both even and odd is $f(x) = 0$. It is also easy to verify this analytically:

> *Suppose that $f(x)$ is both even, so that $f(-x) = f(x)$, and odd, so that $f(-x) = -f(x)$. Then $f(x) = -f(x)$, since both are equal to $f(-x)$, and, adding $f(x)$ to both sides of the last equation, we have $2f(x) = 0$ or $f(x) = 0$.*

14 Graphing a Quadratic Function by Hand

In Exercises 2.4 of the text you are asked in several of the problems to sketch the graph of a quadratic function. As discussed in the text, the graph is a parabola and it is sufficient to plot only two points on the graph, providing one of them is the vertex of the parabola. Obtaining an accurate graph, however, is more difficult when there are no x-intercepts and the y-intercept is very close to the vertex. The following example illustrates a method for dealing with a situation such as this.

EXAMPLE   Find the intercepts and vertex of the parabola determined by the function

$$f(x) = 8x^2 + 4x + \frac{5}{2}.$$

Solution To find the x-intercepts we first determine if $f(x)$ factors easily. We begin by factoring out $\frac{1}{2}$ and noting that a factorization of $f(x)$ must have the form

$$f(x) = \frac{1}{2}(16x^2 + 8x + 5) = \frac{1}{2}(\underline{}x + 1)(\underline{}x + 5).$$

The blanks in the above "factorization" must be one of the pairs 4, 4; 2, 8; 8, 2; 16, 1; or 1, 16. [Note that it is not necessary to consider negative coefficients. Why?] Since none of these pairs works, the expression for $f(x)$ does not easily factor—that is, with integer coefficients after the $\frac{1}{2}$ has been factored out. We next set $f(x) = 8x^2 + 4x + \frac{5}{2} = 0$ and use the quadratic formula. this gives

$$x = \frac{-4 \pm \sqrt{16 - 4(8)(\frac{5}{2})}}{2(8)} = \frac{-4 \pm \sqrt{16 - 80}}{16},$$

which are not real numbers. Thus, the graph of $f(x)$ has no x-intercepts.

To find the y-intercept, which a quadratic function always has, we compute $f(0) = \frac{5}{2}$.

To find the vertex of the parabola we obtain the standard form of the equation by completing the square after first factoring 8 from the two x terms:

$$
\begin{aligned}
f(x) &= 8x^2 + 4x + \frac{5}{2} \\
&= 8\left(x^2 + \frac{1}{2}x + \phantom{\frac{1}{16}}\right) + \frac{5}{2} \\
&= 8\left(x^2 + \frac{1}{2}x + \frac{1}{16}\right) + \frac{5}{2} - 8\left(\frac{1}{16}\right) \\
&= 8\left(x + \frac{1}{4}\right)^2 + 2.
\end{aligned}
$$

In this case, the vertex, $(-\frac{1}{4}, 2)$, and the y-intercept, $(0, \frac{5}{2})$, are close together, making it difficult to sketch an accurate graph by hand. To solve this problem, we simply choose a value for x that is further away from $x = -\frac{1}{4}$, say $x = 1$, and plot the corresponding point on the graph of $f(x)$.

Using $f(1) = 8(1)^2 + 4(1) + \frac{5}{2} = 14.5$ we see that the graph also passes through the point $(1, 14.5)$. [Note that it must also then pass through $(-\frac{1}{4} - \frac{5}{4}, 14.5) = (-\frac{3}{2}, 14.5)$. Why?] The graph of $f(x)$ is shown below to the right of a plot of the vertex and y-intercept.

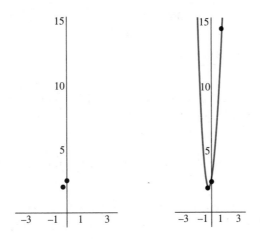

See Section 12 in **Use of a Calculator** (Part II) in this manual for a further discussion on graphing quadratic functions.

$\square$

15 | Domain of a Composition

As discussed in Section 2.6 of the text, the domain of the composition of two functions f and g is the set of all numbers x in the domain of g such that $g(x)$ is in the domain of f. This is somewhat complicated to apply and it is tempting to find the domain of $f \circ g$ by computing a formula for $(f \circ g)(x)$ and then using this formula to determine the domain of $(f \circ g)(x)$. *Be careful;* this method won't always work, as shown in the following example.

■ **EXAMPLE**   Find the domain of $f \circ g$, given

$$f(x) = \frac{3x + 2}{x + 1} \quad \text{and} \quad g(x) = \frac{x}{x - 3}.$$

Solution The domain of g is $\{x \mid x \neq 3\}$ so the domain of $f \circ g$ cannot contain $x = 3$. Now, we must also have $g(x)$ in the domain of f. Since the domain of f is $\{x \mid x \neq -1\}$, we must have $g(x) = x/(x - 3) \neq -1$. Solving $x/(x - 3) = -1$ we have

$$\frac{x}{x - 3} = -1$$
$$x = -1(x - 3) = -x + 3$$
$$2x = 3$$
$$x = \frac{3}{2}.$$

Thus, $x = \frac{3}{2}$ is not in the domain of $f \circ g$. The domain of $f \circ g$ is then $\{x \mid x \neq 3, \frac{3}{2}\}$, or in interval notation, $(-\infty, \frac{3}{2}) \cup (\frac{3}{2}, 3) \cup (3, \infty)$.

$\square$

In this example we could have computed $(f \circ g)(x)$:

$$(f \circ g)(x) = f\big(g(x)\big) = f\left(\frac{x}{x-3}\right) = \frac{3\left(\dfrac{x}{x-3}\right)+2}{\dfrac{x}{x-3}+1} = \frac{3\left(\dfrac{x}{x-3}\right)+2}{\dfrac{x}{x-3}+1}\left(\dfrac{x-3}{x-3}\right)$$

$$= \frac{3x+2(x-3)}{x+(x-3)} = \frac{5x-6}{2x-3}.$$

The formula for $(f \circ g)(x)$ is $(5x-6)/(2x-3)$, which excludes only $x = \frac{3}{2}$. However, as we saw in the previous example, the domain of $f \circ g$ must also exclude $x = 3$. This fact can be lost when using the formula for $f \circ g$ to find the domain of $f \circ g$.

The preceding discussion suggests an alternative method for finding the domain of the composition of two functions.

> *The domain of the composition $f \circ g$ of two functions, f and g, is the intersection of the domains of the formulas for $g(x)$ and $(f \circ g)(x)$.*

This method is especially appropriate when a formula for $f \circ g$ must already be found as part of an exercise. We illustrate this technique with the following example.

■ **EXAMPLE** Find the domain of $f \circ g$, given

$$f(x) = \frac{1}{x^2 - 1} \quad \text{and} \quad g(x) = \sqrt{x+3}.$$

Solution The domain of g is $\{x \mid x+3 \geq 0\} = \{x \mid x \geq -3\}$. We compute

$$(f \circ g)(x) = f\big(g(x)\big) = f\big(\sqrt{x+3}\big) = \frac{1}{\big(\sqrt{x+3}\big)^2 - 1} = \frac{1}{x+3-1} = \frac{1}{x+2}.$$

The domain of the formula for $f \circ g$ is $\{x \mid x \neq -2\}$, so the domain of $f \circ g$ is

$$\{x \mid x \geq -3\} \cap \{x \mid x \neq -2\} = [-3, -2) \cup (-2, \infty).$$

□

16 Synthetic Division

Polynomial long division, when the divisor has the form $x - c$, for any real number c, can be simplified using the technique called **synthetic division**. For example, when $f(x) = 3x^3 + 5x^2 - 4x + 2$ is divided by $d(x) = x + 2$ we see by regular long division

$$
\begin{array}{r}
\boxed{3}\,x^2\ \boxed{-1}\,x\ \boxed{-2} \\
x + \boxed{2}\ \overline{\big)\ \boxed{3}\,x^3 + \boxed{5}\,x^2\,\boxed{-4}\,x + \boxed{2}} \\
\underline{3x^3 + 6x^2} \\
\boxed{-1}\,x^2 - 4x + 2 \\
\underline{-1\,x^2 - 2x} \\
\boxed{-2}\,x + 2 \\
\underline{-2\,x - 4} \\
\boxed{\boxed{6}}
\end{array}
$$

that the quotient is $q(x) = 3x^2 - x - 2$ and the remainder is $r = 6$. We see in this process that much of the writing is extraneous and all of the pertinent information is contained in the boxed numbers, with various styles of boxes used to distinguish the numbers involved in the divisor $\boxed{c}$, the dividend $\boxed{d}$, the quotient $\boxed{q}$, and the remainder $\boxed{\boxed{r}}$. The procedure of synthetic division for dividing $f(x)$, a polynomial of degree $n > 0$, by $x - c$ is summarized as follows.

Synthetic Division

(i) Write c followed by the coefficients of $f(x)$. Be sure to include any coefficients of $f(x)$ that are 0, including the constant term. Note that the form of the divisor is $x - c$, so that if $f(x)$ is divided by $x + 3$, for example, the value of c is -3.

(ii) Bring down the first coefficient of $f(x)$ to the third row.

(iii) Multiply this number by c and write the product directly under the second coefficient of $f(x)$. Then add the two numbers in this column and write the sum beneath them in the third row.

(iv) Multiply this sum by c and write the product in the second row of the next column. Then add the two numbers in this column and write the sum beneath them in the third row.

(v) Repeat the preceding step as many times as necessary.

(vi) The last number in the third row is the constant remainder r; the numbers preceding it in the third row are the coefficients of $q(x)$, the quotient polynomial of degree $n - 1$.

■ **EXAMPLE**   Use synthetic division to find the quotient and remainder when $f(x) = 3x^3 + 5x^2 - 4x + 2$ is divided by $d(x) = x + 2$.

Solution Identifying $c = -2$ we have

$$
\begin{array}{r|rrrr}
-2 & 3 & 5 & -4 & 2 \\
   &   & -6 & 2 & 4 \\
\hline
   & 3 & -1 & -2 & \boxed{6} = r
\end{array}
$$

The coefficients of the quotient are the first three numbers in the third row, so $q(x) = 3x^2 - x - 2$. The remainder is the far right entry in the third row, $r = 6$.

$\square$

As mentioned in the text, one of the applications of synthetic division is the evaluation of a polynomial $f(x)$ at a number c. As the following example shows, this is not always the most effective way to compute $f(c)$.

■ **EXAMPLE**   Evaluate $f(-1)$ if $f(x) = 2x^5 + 3x - 4$.

Solution Since $f(-1)$ is the remainder when $f(x) = 2x^5 + 3x - 4$ is divided by $x + 1$, its value could be found using synthetic division by $x + 1$. In this case, however, it is easier to simply plug -1 into the formula for $f(x)$:

$$f(-1) = 2(-1)^5 + 3(-1) - 4 = -2 - 3 - 4 = -9.$$

Contrasting this with the use of synthetic division

$$
\begin{array}{r|rrrrrr}
-1 & 2 & 0 & 0 & 0 & 3 & -4 \\
   &   & -2 & 2 & -2 & 2 & -5 \\
\hline
   & 2 & -2 & 2 & -2 & 5 & \boxed{-9} = r
\end{array}
$$

which also gives $f(-1) = -9$ (of course), we see that direct evaluation of the function at $x = -1$ involves fewer computations.

$\square$

17   Complex Numbers

Complex numbers naturally arise when considering the solutions of polynomial equations. For example, the simple polynomial equation $x^2 + 1 = 0$ has no real zeros because there is no real number whose square is -1. However, if we define a new number, denoted by i, having the property that $i^2 = -1$, or $i = \sqrt{-1}$, we then see that $x = i$ is a zero of $x^2 + 1$. In general, any expression of the form $z = a + bi$, where a and b are real numbers and $i^2 = -1$, is called a **complex number**. The numbers a and b are called the **real** and **imaginary** parts of z, respectively. A complex number of the form $0 + bi$, $b \neq 0$, is said to be a **pure imaginary number**. Choosing $b = 0$ in $z = a + bi$ we see that every real number is a complex number. Thus, the set of real numbers is a subset of the set of complex numbers.

■ EXAMPLE

(a) The complex number $z = 4 - 5i = 4 + (-5)i$ has real part 4 and imaginary part -5.
(b) $z = 10i = 0 + 10i$ is a pure imaginary number.
(c) $z = -6 + 0i = -6$ is a real number. (It is also a complex number.)

$\square$

17.1   Equality

To define equality of two complex numbers we require that their real parts be equal and their imaginary parts be equal. That is, if $z_1 = a + bi$ and $z_2 = c + di$, then $z_1 = z_2$ if and only if $a = c$ and $b = d$.

17.2   Addition and Subtraction

To add and subtract complex numbers it is helpful to think of them as linear polynomials in the variable i. The complex numbers are then added or subtracted as though they were polynomials.

■ EXAMPLE   If $z_1 = 5 - 6i$ and $z_2 = 2 + 4i$ then find $z_1 + z_2$ and $z_1 - z_2$.

Solution   Thinking of z_1 and z_2 as polynomials and combining like terms we have

$$z_1 + z_2 = (5 - 6i) + (2 + 4i) = (5 + 2) + (-6 + 4)i = 7 - 2i$$

and

$$z_1 - z_2 = (5 - 6i) - (2 + 4i) = (5 - 2) + (-6 - 4)i = 3 - 10i.$$

$\square$

We formally define addition and subtraction of $z_1 = a + bi$ and $z_2 = c + di$ by

$$z_1 + z_2 = (a + bi) + (c + di) = (a + c) + (b + d)i$$

and

$$z_1 - z_2 = (a + bi) - (c + di) = (a - c) + (b - d)i.$$

17.3 | Complex Conjugate |

Consider the quadratic equation $x^2 - 8x + 25 = 0$. Since the left-hand side of this equation does not factor with integer coefficients, we use the quadratic formula and the fact that $\sqrt{-1} = i$ to solve the equation:

$$x = \frac{-(-8) \pm \sqrt{(-8)^2 - 4(1)(25)}}{2(1)} = \frac{8 \pm \sqrt{64 - 100}}{2}$$

$$= \frac{8 \pm \sqrt{-36}}{2} = \frac{8 \pm \sqrt{36}\sqrt{-1}}{2} = \frac{8 \pm 6i}{2} = 4 \pm 3i.$$

In general, whenever the use of the quadratic formula to solve a quadratic equation with real coefficients leads to an expression involving $\sqrt{-1} = i$, say $a + bi$, a second solution will always be $a - bi$. We say that $a - bi$ is the **complex conjugate** of $z = a + bi$ and denote it by $\bar{z} = \overline{a + bi}$. For example, the complex conjugate, or simply **conjugate**, of $z = 2 - 5i$ is

$$\bar{z} = \overline{2 - 5i} = 2 - (-5)i = 2 + 5i.$$

As indicated by the Conjugate Zeros Theorem in the text, if $f(x)$ is a polynomial function of degree $n > 1$, and z is a complex zero of $f(x)$, then the conjugate $\bar{z}$ is also a zero of $f(x)$. This theorem is illustrated in the solution above of the quadratic equation $x^2 - 8x + 25 = 0$. (This is the equation that is solved to find the zeros of the quadratic polynomial function $f(x) = x^2 - 8x + 25$.)

17.4 | Multiplication and Division |

As with addition and subtraction of complex numbers, a formula for the product of two complex numbers $z_1 = a + bi$ and $z_2 = c + di$ can be found by treating the complex numbers as linear polynomials in i and multiplying them using $i^2 = -1$. Rather than memorizing a formula for this, it is better to simply carry out this process each time using the specific complex numbers.

■ **EXAMPLE**   If $z_1 = 5 - 6i$ and $z_2 = 2 + 4i$, find the product $z_1 z_2$.

Solution We use the distributive, commutative, and associative properties. Then

$$z_1 z_2 = (5 - 6i)(2 + 4i) = 5(2) + [5(4) + (-6)(2)]i + (-6)(4)i^2$$
$$= 10 + (20 - 12)i - 24(-1) = 34 + 8i.$$

$\square$

To divide complex numbers and represent the quotient in the form $a + bi$ we use the fact that the product of a complex number and its conjugate is a real number. To see this let $z = a + bi$ and multiply:

$$z\bar{z} = (a + bi)(a - bi) = a^2 - (bi)^2 = a^2 - b^2 i^2 = a^2 - b^2(-1) = a^2 + b^2.$$

The following example illustrates how to find the quotient of two complex numbers.

■ **EXAMPLE**   If $z_1 = 3 - 2i$ and $z_2 = 4 + 5i$, find z_1/z_2.

Solution To compute the quotient we multiply both the numerator and denominator of z_1/z_2 by the conjugate of z_2:

$$\frac{z_1}{z_2} = \frac{3 - 2i}{4 + 5i} = \frac{3 - 2i}{4 + 5i}\left(\frac{4 - 5i}{4 - 5i}\right) = \frac{12 + [3(-5) + (-2)(4)]i + (-2)(-5)i^2}{4^2 + 5^2}$$

$$= \frac{12 - 23i + 10(-1)}{16 + 25} = \frac{2 - 23i}{41} = \frac{2}{41} - \frac{23}{41}i.$$

$\square$

18 | Checking Your Answer

When solving a problem, especially one that involves a numerical solution, it is *always* a good idea to check your result. This can generally be done in a variety of ways, a few of which are discussed here.

Rework the Problem. This is perhaps the most obvious way to check your result. Two fairly obvious drawbacks, however, are that it can be time consuming, and that an error made the first time through the problem may well be repeated the second time.

Reasonableness. Is the answer at least about what you expected it to be, or does it make sense as a solution? For example, if you want to find the x-intercepts of the graph of $y = 2x^2 - x - 7$ and you conclude from the quadratic formula that there are none, you can check this result by noting that the discriminant, $(-1)^2 - 4(2)(-7) = 57$, is positive and therefore the equation $y = 2x^2 - x - 7$ has real solutions. Thus, the graph of $y = 2x^2 - x - 7$ must, in fact, have two distinct x-intercepts, and your original contention that there are no x-intercepts must be incorrect.

Another example of an unreasonable answer is a very small or very large numerical result when you expect the answer to be something in the range $(20, 100)$. Or maybe you get a negative result when you know that the answer must be positive. Sometimes a calculator can be used to convert an exact value like $1/(\sqrt{95} - 3\pi)$ into a decimal approximation that can easily be seen to be positive or negative, or large or small. While the reasonableness criterion will not give you any idea of what the actual answer is, it is very quick and easy to apply.

Draw a Graph. The problem may not require a graph, but it is generally easy enough to obtain the graph of a function and at least determine if your answer is close to the answer indicated by the calculator.

For example, consider the trigonometric problem of finding all values of x in the interval $[0, 2\pi]$ for which $\sin x = \cot x$. We use $\cot x = \cos x / \sin x$, the Pythagorean identity $\sin^2 x + \cos^2 x = 1$, and solve for $\cos x$:

$$\sin x = \cot x$$
$$\sin x = \frac{\cos x}{\sin x}$$
$$\sin^2 x = \cos x$$
$$1 - \cos^2 x = \cos x$$
$$\cos^2 x + \cos x - 1 = 0$$
$$\cos x = \frac{-1 \pm \sqrt{1 + 4}}{2}$$
$$\cos x = -\frac{1}{2} \pm \frac{1}{2}\sqrt{5}.$$

Since $-\frac{1}{2} - \frac{1}{2}\sqrt{5} < -1$, we have only $\cos x = -\frac{1}{2} + \frac{1}{2}\sqrt{5}$ or $x = \cos^{-1}(\frac{1}{2}\sqrt{5} - \frac{1}{2})$, and since $\cos x$ is positive in the 4th quadrant, $x = 2\pi - \cos^{-1}(\frac{1}{2}\sqrt{5} - \frac{1}{2})$. Using a calculator or computer, we have $x = 0.904557$ and $x = 5.37863$. These results can be checked by plotting $y = \sin x$ and $y = \cot x$ on the same set of axes and observing that they do in fact intersect at about $x = 0.9$ and $x = 5.4$. While this does not absolutely guarantee that our answer is exactly correct, it does show that at least we are very close.

19 |Extraneous Roots of Logarithmic Equations|

When solving a logarithmic equation, it is possible to obtain an extraneous solution. This means that it is essential that you check your answers when solving an equation involving logarithms.

■ **EXAMPLE**   Solve $\ln x + \ln(x + 3) = \ln(2x)$.

Solution   Using the properties of logarithms and the fact that $\ln x$ is a one-to-one function, we have

$$\ln x + \ln(x + 3) = \ln 2x$$
$$\ln x(x + 3) = \ln 2x$$
$$x(x + 3) = 2x$$
$$x^2 + 3x = 2x$$
$$x^2 + x = 0$$
$$x(x + 1) = 0.$$

Thus, $x = 0$ or $x = -1$. When we check this result, however, we note that the original equation contains $\ln x$, so x must be positive. Thus, neither of our "solutions" checks, and we conclude that the equation has no solutions.   □

PART II

Use of a Calculator

Use of a Calculator

1 | Introduction |

This part is not intended to be a comprehensive manual on the use of a graphing calculator in a precalculus course. While much of the material discussed will be pertinent to any graphing calculator, the references in this manual will be to the TI-84 family of calculators. A number of students entering college already have a TI-83 calculator that they used in their high school mathematics classes. The TI-83 is similar to the TI-84 and most of what is discussed here applies to the TI-83 as well. After a few brief comments on the use of the TI-84 calculator, the focus in this manual will be on how to use the calculator to assist in the solution of a number of the problems in the text. While the TI-84 calculator contains many sophisticated routines that could be used, the focus in this manual will be on using techniques that require a minimum amount of expertise with the calculator. Before deciding on the extent to which you use a calculator in your precalculus course, it is very important that you keep two things in mind:

1. Your instructor may not allow the use of calculators on quizzes or tests. If this is the case you should definitely avoid becoming dependent on the calculator as a source of solutions to problems. At most, you should use the calculator as a way to check the plausibility of your answer to a problem.

2. Even if you are allowed free use of a calculator it is important to keep in mind that many of the problems in the text are frequently designed to have "nice" outcomes. That is, the answers will generally involve integers or relatively simple fractions. In actual practice, this will not generally be the case, and your calculator will give only

an approximate answer. You may still need to generate the exact answer (for example, in a calculus course following this precalculus course). A simple example involves the equation $x^2 - 3 = 0$. The exact solution is $x = \sqrt{3}$ or $x = -\sqrt{3}$, but the calculator will generate the solutions $x = 1.7320508075\ldots$ and $x = -1.732050807\ldots$, which are very close, but still not exact.

2   Mode Settings

When you first turn on the calculator it is a good idea to check the settings in the calculator. To do this press the MODE key to see something like

Generally, you will want to use the default settings, which are highlighted in the screenshot shown above.

3   Clearing the Calculator Screen

To clear the screen when you are not in a graphics mode simply position the cursor on a blank line and press the CLEAR key. If the cursor is not on a blank line, pressing the CLEAR key will generally erase only that line. In this case, if there are still entries on the screen and you want to erase them, press the CLEAR key again.

When in the Y= window, to clear a function definition, position the cursor at the end of the function and press the CLEAR key.

Suppose a graphics object, like a line connecting $(1, 3)$ and $(-5, -4)$, has been entered using the DRAW menu. Then, later, you want to graph the function $y = x^2$. In this case you are likely to see the following screen:

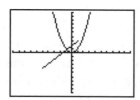

You probably didn't expect to see the line in the graph. To clear the line, press the 2ND DRAW keys and select the first option, 1 : ClrDraw

When the CLEAR key doesn't work, try using the 2ND QUIT keys to clear a graphics screen.

4 Basic Calculations and Memory

In the TI-84 calculator, expressions like $(2\pi - 3)(5 + \sqrt{2})^3$ can be computed as shown:

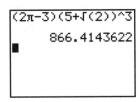

If the result of this operation is to be used again, *do not* write the number 866.4143622 on a piece of paper and then key it in (or even worse, a rounded version of it like 866.41) at the appropriate spot. For example, if you later want to find the reciprocal of the square root of this number, then you should store it in a memory register and call it up when needed. This process is demonstrated here.

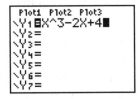

The fourth line in the window, which stores the result of the calculation in the variable B, is obtained by pressing the STO ALPHA B keys. To clear a value assigned to a variable, simply store the number 0 in that variable.

5 Functions

To enter a function that is to be evaluated or graphed, press the Y= key and enter the function (or functions) that is to be used. In the screenshot below, the function $y = x^3 - 2x + 4$ is entered.

5.1 Function Evaluation

If a function has been entered as Y1 then it can be evaluated at any number, say $x = -3.2$, using the 2ND TBLSET and 2ND TABLE keys. The following sequence of screenshots demonstrates how to do this.

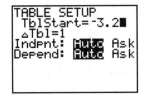 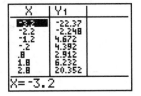

In this case, we see that when $x = -3.2$, $x^3 - 2x + 4$ equals -22.37. [Note: When entering a negative number like -3.2 be sure to use the $\boxed{(-)}$ key in the bottom row on the calculator, as opposed to the subtraction key $\boxed{-}$ just to the right of the $\boxed{6}$ key. *This is a very common mistake, and should generally be the first possibility you consider whenever the calculator gives you an error message.* Also, once you have input the number in the `TABLESETUP` screen, be sure to press the $\boxed{\text{ENTER}}$ key.]

5.2 $\boxed{\text{Graphing a Function}}$

To graph `Y1` use the $\boxed{\text{WINDOW}}$ key to set the viewing rectangle. For example, if you set Xmin=-10, Xmax=10, Ymin=-10, and Ymax=10, and then press the $\boxed{\text{GRAPH}}$ key, you obtain the following two screens:

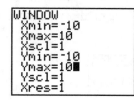 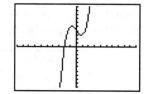

To change the viewing rectangle simply press the $\boxed{\text{WINDOW}}$ key and reset the boundaries of the window. Based on the screen shown above, you might want to reset Xmin to be -4 and Xmax to be 3. In this case the following screen is obtained:

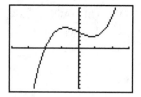

5.3 $\boxed{\text{An Alternative Method of Function Evaluation}}$

An alternative way to evaluate a function that has been entered in `Y1` uses the $\boxed{\text{2ND}}$ $\boxed{\text{CALC}}$ key combination. Suppose you want to evaluate $f(-3.4)$ when

$$f(x) = \frac{\sqrt{x^2 - 3}}{x + 4}.$$

First, enter this function as `Y1` in the $\boxed{\text{Y=}}$ window. Next, be sure that the graphing window includes the value -3.4. Then, in the $\boxed{\text{2ND}}$ $\boxed{\text{CALC}}$ window, choose `1 : value`. Enter the value for x, in this case -3.4, and press $\boxed{\text{ENTER}}$. (Did you remember to use the $\boxed{(-)}$ key instead of the subtraction key?) In the lower right hand portion of the screen you will see the value of $f(x)$, in this case `Y = 4.8762463`. You should see something like the following sequence of screens:

  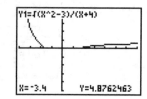

6   Solution of an Equation

To find the solution or solutions of an equation such as $x^3 - x^2 - 12x + 7 = 0$, first graph the function to get a sense of how many solutions there are and approximately where they are. The graph of $\texttt{Y2} = x^3 - x^2 - 12x + 7 = 0$ is shown below with $\texttt{Xmin} = -10, \texttt{Xmax} = 10,$ $\texttt{Ymin} = -10,$ and $\texttt{Ymax} = 10.$

From the graph above, we see that the equation probably has three solutions near $x = -3$, $x = 1$, and $x = 4$. The word "probably" is used because, without knowledge of the behavior of cubic functions, it is possible that the graph comes back down (or up on the left) and crosses the x-axis outside the window of the calculator. Now, to find the solutions of the equation, use the EQUATION SOLVER in the the MATH menu. This can be accessed by pressing the $\boxed{\text{MATH}}$ key followed by the $\boxed{0}$ key. Where the screen reads $\texttt{eqn}: 0 =$ type in the equation. (If the screen does not say **EQUATION SOLVER** at the top, press the up cursor.) You should see the following screen:

Now press $\boxed{\text{ENTER}}$ and assign a value to $\texttt{X}$ that is close to a solution. For example, to find the smallest positive solution of the equation in this example you could assign $\texttt{x} = 1$ followed by $\boxed{\text{ALPHA}}$ $\boxed{\text{SOLVE}}$. You should see the following sequence of two screens:

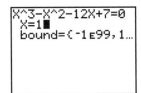 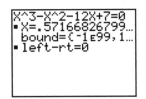

The smallest positive solution is $x = .57166826799$ rounded to 10 decimal places. To find the other positive solution press $\boxed{4}$ followed by $\boxed{\text{SOLVE}}$.

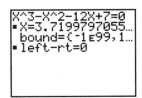

This shows that the next positive root is $x = 3.7199797055$.

7 | Solution of an Inequality

To find the solution set of an inequality such as $x^2+3x-10 \geq 0$, first graph $\mathtt{Y1} = x^2+3x-10$ and then solve the equation $x^2 + 3x - 10 = 0$ as described in **Solution of an Equation** on the previous page. This gives $x = -5$ and $x = 2$. Now, pressing the $\boxed{\text{GRAPH}}$ key again shows that $x^2 + 3x - 10 \geq 0$ (that is, above the x-axis) when $x \leq -5$ and $x \geq 2$. You should see the following sequence of calculator screens:

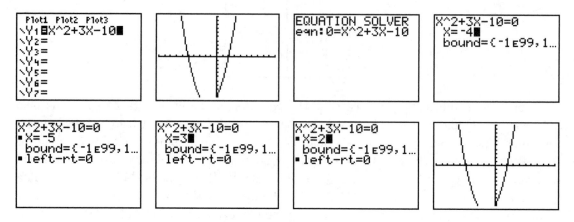

While the above technique could be used to approximate the solution of an inequality, it should only be used to check the answer that you arrived at using the sign chart technique described in the text.

8 | Solution of an Inequality Involving an Absolute Value

To find an approximation to the solution of an inequality involving an absolute value such as $|2x - 8| \geq 12$, first write the inequality as $|2x - 8| - 12 \geq 0$. Then graph $\mathtt{Y1} = \mathtt{abs}(2x - 8) - 12$ and find the solutions of $|2x - 8| = 12$ by estimating where the graph intersects the x-axis. The solution set of the inequality will then be the union of the intervals where the graph is above the x-axis.

9 | Graphing a Circle

Option 9 in the DRAW menu is used to draw a circle. In this case, three inputs are needed—the first two are the coordinates of the center and the third is the radius of the circle. For example, to draw a circle centered at $(1, 2)$ with radius 3 use the following sequence of keystrokes: $\boxed{\text{2ND}}$ $\boxed{\text{DRAW}}$ $\boxed{9}$ $\boxed{1}$ $\boxed{2}$ $\boxed{3}$. You will probably see a screen that looks something like the following:

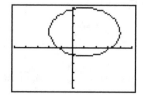

To make this circle actually look like a circle rather than an oval, use the ZOOM key followed by 5 : ZSquare. To see the resulting circle you may need to press the CLEAR key followed by the ENTER key. The screen should now show the following:

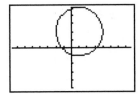

10   Limit of a Function

While you should generally not use your calculator to find the limit of a function, it can be useful to check that the limit you have obtained using hand computations is plausible. For example, suppose you want find

$$\lim_{x \to -1} \frac{2x + 2}{x^2 - x - 2}.$$

By hand, you compute

$$\frac{2x + 2}{x^2 - x - 2} = \frac{2(x + 1)}{(x - 2)(x + 1)} = \frac{2}{x - 2}, \quad x \neq -1,$$

so that

$$\lim_{x \to -1} \frac{2x + 2}{x^2 - x - 2} = \lim_{x \to -1} \frac{2}{x - 2} = \frac{2}{-1 - 2} = -\frac{2}{3}.$$

There are two ways you can use your calculator to check the plausibility of this result. First, graph the function

$$Y1 = \frac{2x + 2}{x^2 - x - 2},$$

and then use the TRACE key to position the cursor over $x = -1$ (or as close as you can get to $x = -1$). The corresponding y-coordinate will be displayed in the lower right-hand portion of the screen, as shown here:

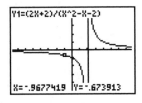

Since Y $= -.673913$ is close to $-\frac{2}{3}$, you have reason to believe that your hand computations are correct.

Alternatively, you could evaluate the function (as discussed in Section 5.1 of this part of the manual)

$$Y1 = \frac{2x + 2}{x^2 - x - 2},$$

at a value very close to $x = -1$, say, at $x = -1.01$. In this case, you will see a screen like the one here:

In this case, when $x = -1.01$ the corresponding y value is $-.6645$. Since this number is close to $-\frac{2}{3}$, you have reason to believe that your hand computations are correct.

11   Factorials

Factorials play a very important role in probability, which explains its location in the calculator. To find, say 6!, press 6 MATH. You will see the following screen:

Now press the right cursor three times to highlight PRB followed by 4 ENTER. This will give the value of 6!.

To test your ability with factorials on the calculator, you might want to find values of n for which the calculator switches to scientific notation, and values for which it overflows. You should also determine what the calculator thinks 0! is, and what it thinks of $(-4)!$. [In the latter case, be sure to include the parentheses around the -4 before trying to find the factorial.]

12   Graphing Quadratic Functions

As discussed in Section 2.4 of the text, the graph of a quadratic function is always a parabola opening upward or downward. When the coefficients of the function are integers or simple fractions you should be able to sketch a graph of the function by hand. In this case, a calculator may be used to confirm the results of your work. However, when the function does not have simple coefficients, it can be very tedious to find the vertex and the intercepts by hand.

Suppose, for example, that you want to approximate the intercepts and vertex of the parabola determined by $f(x) = 2.83x^2 - 4.32x - 31.09$. To get a sense of the intercepts and vertex of the parabola you could graph the function. To do this, enter the function as Y1 using the Y= key and make sure that all other functions, Y1, Y2, Y3, ..., are clear.

Then press the ⌈GRAPH⌉ key. Depending on the settings in the ⌈WINDOW⌉ screen you will see a screen like

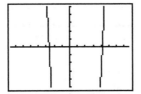

Knowing that the graph is a parabola, you can zoom in on the x-intercepts to determine their locations with reasonable accuracy. To do this, first press the ⌈TRACE⌉ key and then use the left and right cursor keys, ⌈◄⌉ and ⌈►⌉, to place the cursor near an x-intercept. (Initially, the cursor may be off the screen, but its x- and y-coordinates are shown at the bottom of the screen, so you can move it into the viewing window.) Having done this, you should see a screen like the following:

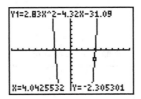

Now, press the ⌈ZOOM⌉ ⌈2⌉ ⌈ENTER⌉ key sequence to zoom in on the part of the graph around the cursor. At this point, the cursor is no longer attached to the graph and can be moved using any of the four cursor keys, making it easier to position the cursor more directly on top of the x-intercept. After doing this, press the ⌈ZOOM⌉ ⌈2⌉ ⌈ENTER⌉ key sequence to zoom in again. Successively repositioning the cursor and noting the x-values (the y-value should remain 0), you soon see that, to two decimal places, this x-intercept is 4.16.

To zoom back out to find the other x-intercept you can either use the ⌈ZOOM⌉ ⌈3⌉ ⌈ENTER⌉ key sequence, or you can reset Xmin, Xmax, Ymin, and Ymax in the ⌈WINDOW⌉ screen. For example, if you set Xmin $= -10$, Xmax $= 10$, Ymin $= -10$, and Ymax $= 10$ and then press the ⌈GRAPH⌉ key you will see the following screen:

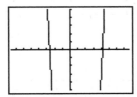

Repeating the process above to find the other x-intercept, you should determine that it is approximately $x = 2.637$. (If in the above process you lose the y-axis from the screen, it can always be recovered by resetting the Ymin and Ymax values in the ⌈WINDOW⌉ menu.)

Notice in the picture above (where Xmin $= -10$, Xmax $= 10$, Ymin $= -10$, and Ymax $= 10$) that both the vertex and the y-intercept of the parabola are below the bottom of the calculator screen. To get a more complete view of the parabola, you could change Ymin using the [WINDOW] menu to, say, Ymin $= -20$. This is better, but still not enough, so try again. Eventually, Ymin $= -40$ will work and you should see the following screen:

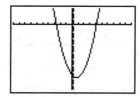

At this point, you can see the y-intercept and could approximate its coordinates using the [TRACE] and [ZOOM] keys as above when approximating the x-coordinates, but this is an overuse of the graphing capabilities of the calculator. Instead, simply use one of the techniques described earlier in Section 5.1 or 5.3 to find $f(0)$. In this case, it is easier to use Section 5.3, so in the [2ND] [CALC] window choose 1 : value. Enter the value for x, 0 for the y-intercept, and press [ENTER]. In the lower right hand portion of the screen you will see the value of $f(x)$, in this case Y $= -31.09$. You should see something like the following screen:

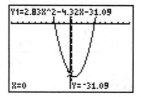

Next, to find the vertex of the parabola, you can use the [TRACE] and [ZOOM] keys as described above for finding the x-intercepts. You should find that the vertex is approximately at $(0.76, -32.74)$.

13   [Constructing a Table]

The table building capability of the calculator can be used to generate the data in a table. We illustrate this by considering part (d) of Problem 51 in Section 2.4 of the text. We begin by entering the function $R(D) = kD(P - D)$ as Y1 in the [Y=] window. To do this we let $k = 0.00003$, $P = 10,000$, and identify D with X. You should see the following screen:

```
Plot1  Plot2  Plot3
\Y1■.00003*X*(10
000-X)■
\Y2=
\Y3=
\Y4=
\Y5=
\Y6=
```

Next, use the [2ND] [TBLSET] key combination to change the Indpnt : setting from Auto to Ask. (The TblStart and ΔTbl setting do not matter.) You should now see the following window with `Ask` flashing after you press the $\boxed{\text{ENTER}}$ key:

Now press the $\boxed{\text{2ND}}$ $\boxed{\text{TABLE}}$ key combination and enter 125 after `X =` at the bottom of the window, followed by the $\boxed{\text{ENTER}}$ key. You should see 125 in the `X` column of the table and 37.031 in the `Y1` column of the table. Round 37.031 to the nearest integer since it represents a number of persons, add it to 125, and enter the result 162 after `X =`. Press the $\boxed{\text{ENTER}}$ key, and you should see

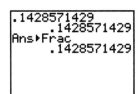

Continuing in this fashion, you will generate the following table containing the number of infected individuals in the `Y1` column:

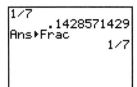

14 $\boxed{\text{Converting a Decimal to a Fraction}}$

Selecting 1:▶Frac in the $\boxed{\text{MATH}}$ menu will sometimes convert a decimal to a fraction. The following screens illustrate this:

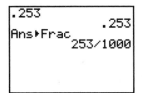 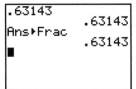

Notes: (1) After the decimal number is entered, press the $\boxed{\text{MATH}}$ key, followed by $\boxed{1}$ $\boxed{\text{ENTER}}$ to obtain the fraction.

(2) As the second screen shows, not every decimal can be converted to a fraction. You can try experimenting with various decimals to see if you can determine those that will be converted to fractions.

(3) If the decimal arises as the quotient of two integers, as shown above in the third screen, the decimal result *may (but not always)* be converted back to the fraction. On the other hand, when the same decimal is entered directly, it may not be converted to a fraction. See the fourth screen above.

15   Arithmetic of Complex Numbers

It is a relatively simple task to have the calculator perform the four basic arithmetic operations of addition, subtraction, multiplication, and division on complex numbers. To do this, you first need to put the calculator in rectangular complex mode. Press the MODE key, scroll down to REAL, and then scroll right to $a + bi$. Then press the ENTER key.

To find $(2-4i)-(1+3i)$, first clear the calculator screen (see Section 3 in this manual); then the following sequence of keystrokes will cause the calculator to display the difference of the two complex numbers: (2 − 4 2ND i) − (1 + 3 2ND i) ENTER .

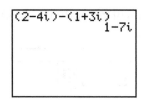

A quotient of complex numbers with integer real and imaginary parts will usually result in a complex number with fractional (meaning noninteger) real and imaginary parts. The calculator will display the resulting quotient as a complex number with decimal real and imaginary parts. In this case, decimal to fraction conversion (see Section 14 of this manual) can be used. The following screen illustrates this in the computation of $(5 + 7i)/(4 - 2i)$:

16   Rational Zeros of a Polynomial With Integer Coefficients

We want to find the rational zeros of a polynomial $f(x) = a_n x^n + a_{n-1} x^{n-1} + \cdots + a_1 x + a_0$, where $n \geq 1$ and a_i is an integer for $0 \leq i \leq n$. As discussed in the text, all rational zeros of $f(x)$ must have the form p/s where p is an integer factor of a_0 and s is an integer factor of a_n. With the calculator it is a simple matter to directly check each possible rational zero. Using the $\boxed{\text{Y=}}$ key enter the function as Y1. In the screen below we are trying to find the rational zeros of $f(x) = 6x^4 + 11x^3 + 14x^2 - 7x - 6$.

Now press the $\boxed{\text{TRACE}}$ key. The set of possible rational zeros is ± 1, $\pm\frac{1}{2}$, $\pm\frac{1}{3}$, $\pm\frac{1}{6}$, ± 2, $\pm\frac{2}{3}$, ± 3, $\pm\frac{3}{2}$, and ± 6. To test -1 simply press $\boxed{(-)}$ $\boxed{1}$ $\boxed{\text{ENTER}}$ and read Y = 10. Thus, $f(1) = 10$, so 1 is not a zero of $f(x)$. Next, press $\boxed{1}$ $\boxed{\text{ENTER}}$ and read Y = 18. Continuing in this manner we find that $-\frac{1}{2}$ and $\frac{2}{3}$ are the rational zeros of $f(x)$. Some sample screens are shown here:

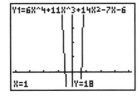 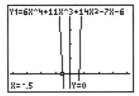  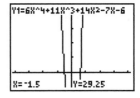

You should be aware that the technique described above is very mechanical and will easily find the rational zeros of $f(x)$. It will not, however, find any irrational zeros, nor will it find any nonlinear factors of $f(x)$. Most significantly, the technique will not contribute to your understanding of the concepts involved in the theory of rational zeros of polynomials.

17   Calculation of the Other Trigonometric Functions

The keyboard of the calculator contains no keys for directly computing the cotangent, secant, cosecant, inverse cotangent, inverse secant, and inverse cosecant functions. To compute the cot, sec, and csc functions, we use the identities

$$\cot x = \frac{1}{\tan x}, \quad \sec x = \frac{1}{\cos x}, \quad \text{and} \quad \csc x = \frac{1}{\sin x}.$$

For example, to find $\cot 137°$, first be sure that the calculator is set to degree mode by pressing $\boxed{\text{MODE}}$, then scroll down to degree and press $\boxed{\text{ENTER}}$ to select degree mode:

After using ⌜CLEAR⌝ to return to the main window, use the keystrokes ⌜TAN⌝ ⌜1⌝ ⌜3⌝ ⌜7⌝ ⌜)⌝ ⌜X⁻¹⌝ to see that cot 137° = −1.07236871. The screen output is shown here.

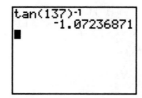

Alternatively, you could use ⌜1⌝ ⌜÷⌝ ⌜TAN⌝ ⌜1⌝ ⌜3⌝ ⌜7⌝ ⌜)⌝ to obtain the same result.

To compute the arccotangent, arcsecant, and arccosecant functions, use the identities

$$\operatorname{arccot} x = \frac{\pi}{2} - \arctan x, \quad \operatorname{arcsec} x = \arccos\left(\frac{1}{x}\right), \quad \text{and} \quad \operatorname{arccsc} x = \arcsin\left(\frac{1}{x}\right)$$

from Problems 64, 65, and 66 in Section 4.7 of the text.

PART III

Basic Skills

Inequalities, Equations, and Graphs

1

Introduction Each section in the text contains problems that focus primarily on manipulative skills. This part of the manual lists those skills. It does not make reference to the other types of problems contained in most sections of the text that require a deeper conceptual understanding of the material.

1.1 The Real Line

- Solve a linear inequality using the properties of inequalities

- Solve a nonlinear inequality using the sign chart method

- Express the solution of an inequality in interval notation

- Graph the solution set of an inequality

- Solve word problems involving inequalities

1.2 Absolute Value

- Find the absolute value of an expression containing an unknown number, x, subject to a constraint on x

- Find the distance between two numbers on the real line

- Find the midpoint of two numbers on the real line

- Solve an equation involving an absolute value

- Solve an inequality involving an absolute value

- Find an inequality involving an absolute value having a given interval or union of intervals as its solution set

1.3 Rectangular Coordinate System

- Locate the quadrant in which a point lies based on the algebraic signs of its coordinates

- Find the distance between two points in the plane

- Find the midpoint of the line segment joining two points in the plane

1.4 Circles and Graphs

- Given the equation of a circle, find its center and radius (completing the square if necessary)

- Determine which of the four possible semicircles is defined by a given equation solved for x or y

- Find the x- and y-intercepts of an equation

- Determine if the graph of an equation possesses symmetry with respect to the x-axis, the y-axis, or the origin

1.5 Calculus Preview—Algebra and Limits

- Simplify an algebraic expression by factoring and adding rational expressions

- Find the limit of an algebraic expression by simpifying it

2 Functions

2.1 Functions and Graphs

- Find the domain of a function given a formula for the function

- Find the domain of a function given the graph of the function

- Find the value of a function at a point in its domain given a formula for the function

- Find the value of a function at a point in its domain given the graph of the function

- Use the vertical line test to determine if a given graph is the graph of a function

- Find the zeros of a given function

- Find the x-intercepts of a given function

- Find the y-intercept of a given function

- Evaluate the factorial function at a positive integer

2.2   Symmetry and Transformations

- Determine if a function is even or odd

- Given a partial graph of a function that is known to be even or odd, find the entire graph

- Given the graph of a function, sketch the graph of a translation, reflection, stretching, or compression of the function

2.3   Linear Functions

- Find the slope of the line through two given points

- Given the equation of a line, find the slope and intercepts of the line

- Find the equation of the line through a given point with a given slope

- Find the equation of the line through two given points

- Find the equation of a line that is parallel to a given line

- Find the equation of a line that is perpendicular to a given line

- Determine if two given lines are parallel to each other

- Determine if two given lines are perpendicular to each other

- Find the point of intersection of two given lines

2.4   Quadratic Functions

- Complete the square to express a quadratic function in standard form

- Find the vertex of the graph of a quadratic function

- Find the x- and y-intercepts of the graph of a quadratic function

- Find the maximum or minimum value of a quadratic function

- Identify the transformations on the graph of $y = x^2$ used to obtain the graph of a quadratic function

2.5 Piecewise-Defined Functions

- Evaluate a piecewise-defined function at a given number

- Find the intercepts of a piecewise-defined function

- Sketch the graph of a piecewise-defined function

- Sketch the graph of functions defined using the greatest integer function

2.6 Combining Functions

- Given formulas for $f(x)$ and $g(x)$, find formulas for $(f+g)(x), (f-g)(x), (fg)(x)$, and $(f/g)(x)$

- Find the composition of two or more functions

- Find the domain of the composition of two functions

- Given a function F that is the composition of functions f and g, find $f(x)$ and $g(x)$

2.7 Inverse Functions

- Use the horizontal line test to determine if a function is one-to-one

- Use the definition to show that a given function is, or is not, one-to-one

- When appropriate, show that a function $f(x)$ is not one-to-one by finding $x_1 \neq x_2$ such that $f(x_1) = f(x_2)$

- Given the domain and range of a one-to-one function, find the domain and range of its inverse function

- Given a formula for a one-to-one function, find a formula for its inverse function

- Given the graph of a one-to-one function, sketch the graph of its inverse function

2.8 Translating Words into Functions

- Find the objective function associated with a word problem

2.9 | Calculus Preview—The Tangent Line Problem

- Compute the difference quotient, $[f(a+h) - f(a)]/h$, for a given function at a given value of a

- Find the equation of the tangent line to a given function at a given point

- Compute the derivative of a given function

- Use the derivative of a given function to compute the slope of the tangent line to the graph of the function at a given x-coordinate on the graph of the function

- Compute the difference quotient, $[f(x) - f(a)]/(x - a)$, for a given function

- Use the difference quotient, $[f(x) - f(a)]/(x - a)$, to find the derivative of $f(x)$ and $x = a$

Polynomial and Rational Functions

3.1 Polynomial Functions

- Determine if a polynomial function is even, odd, or neither

- Graph a polynomial function using its zeros and degree

3.2 Division of Polynomial Functions

- Use long division to find the quotient and remainder of two polynomial functions

- Use the Remainder Theorem to find the remainder when a polynomial is divided by a linear polynomial

- Use polynomial long division to find the value of a polynomial at a number

- Use synthetic division to find the quotient and remainder when a polynomial is divided by a linear polynomial

- Use synthetic division and the Remainder Theorem to evaluate a polynomial function at a given number

3.3   Zeros and Factors of Polynomial Functions

- Find the zeros of a polynomial function and give the complete factorization of the polynomial

- Determine if a given number is a zero of a polynomial

- Use synthetic division to determine if a given linear polynomial is a factor of a given polynomial

- Use polynomial long division to determine if a given polynomial is a factor of another polynomial

- Determine if a given complex number is a zero of a given polynomial

- Given a complex zero of a polynomial, find all other zeros and completely factor the polynomial

- Given the zeros, with their multiplicities, of a polynomial function, find the polynomial function

- Completely factor a polynomial to determine the zeros and their multiplicities of the polynomial

3.4   Real Zeros of Polynomial Functions

- Determine all *possible* rational zeros of a polynomial function having integer coefficients

- Determine all rational zeros of a polynomial function having integer coefficients

- Find all real zeros of a polynomial function with rational coefficients (when possible) and then factor the polynomial using only real numbers

3.5   Rational Functions

- Find the vertical asymptotes of a rational function

- Find the horizontal asymptote of a rational function

- Find the slant asymptote of a rational function whose numerator has degree one greater than the degree of the denominator

- Determine where the graph of a rational function crosses any horizontal or slant asymptotes

- Find any holes in the graph of a rational function

- Graph a rational function

3.6 Partial Fractions

- Find the partial fraction decomposition of a rational function whose numerator has smaller degree than that of the denominator

- Find the partial fraction decomposition of a rational function whose numerator has equal or larger degree than that of the denominator

3.7 Calculus Preview—The Area Problem

- Assuming that the graph of a function $f(x)$ is never negative on an interval $[a, b]$, approximate the area bounded by the lines $x = a$, $x = b$, the graph of the function, and the x-axis

4 Trigonometric Functions

4.1 Angles and Their Measurement

- Draw an angle given in either radian or degree form in standard form

- Convert an angle given in radians to degrees

- Convert an angle given in degrees to radians

- Find an angle between 0° and 360° that is coterminal with a given angle

- Find an angle between 0 and 2π radians that is coterminal with a given angle

- Find angles (when possible) that are complementary and supplementary to a given angle

- Find the arc length subtended by a given central angle in a circle with a given radius

4.2 The Sine and Cosine Functions

- Given the sine of an angle t in a specified quadrant, find the cosine of the angle

- Given the cosine of an angle t in a specified quadrant, find the sine of the angle

- Find the reference angle of a given angle

- Find the exact values of the sine and cosine of all angles having reference angle 0, $\pi/6$, $\pi/4$, $\pi/3$, or $\pi/2$

4.3   Graphs of Sine and Cosine Functions

- Sketch one cycle of the graph of a function involving $\sin x$ or $\cos x$ using shifting, stretching, compressing, and reflecting

- Given one cycle of a sine or cosine graph having period 2π, match it with an equation of the form $y = A\sin x + D$ or $y = A\cos x + D$

- Find the x-intercepts of a sine or cosine graph

- Given one cycle of a sine or cosine graph, match it with an equation of the form $y = A\sin x$ or $y = A\cos x$

- Find the amplitude, period, and phase shift of a function of the form $y = A\sin(Bx + C)$ or $y = A\cos(Bx + C)$

4.4   Other Trigonometric Functions

- Find the exact value of any of the six trigonometric functions of an angle with reference angle 0, $\pi/6$, $\pi/4$, $\pi/3$, or $\pi/2$

- Given the value of any of the six trigonometric functions of an angle t in a specified quadrant, find the values of the other five trigonometric functions of that angle

- Find the period, x-intercepts, and vertical asymptotes of any of the four functions of the forms $y = A\tan(Bx + C)$, $y = A\cot(Bx + C)$, $y = A\sec(Bx + C)$, and $y = A\csc(Bx + C)$

4.5   Special Identities

- Use a sum or difference formula to find the exact value of the sine, cosine, or tangent of a given angle

- Use a double-angle formula to simplify an expression

- Use a half-angle formula to find the exact value of the sine, cosine, or tangent of a given angle

- Reduce a given expression to one of the form $A\sin(Bx + \phi)$

4.6   Trigonometric Equations

- Find all solutions of a given trigonometric equation in which the unknown represents a real number

- Find all solutions of a given trigonometric equation in which the unknown represents an angle measured in degrees

- Find the first three positive x-intercepts of a function involving trigonometric functions

4.7   Inverse Trigonometric Functions

- Find the value of an inverse trigonometric function of a real number

- Find the value of a trigonometric function of an inverse trigonometric function

- Find the value of an inverse trigonometric function of a trigonometric function

- Express $f(g^{-1}(x))$, where f is one of the six basic trigonometric functions and g is one of the six basic inverse trigonometric functions, in terms of x without any trigonometric functions

- Sketch the graphs of the six basic inverse trigonometric functions

- Use inverse trigonometric functions to solve a trigonometric equation

4.8   Right Triangle Trigonometry

- Given two sides of a right triangle, find the six trigonometric functions of an angle of the triangle

- Given two sides or one angle and one side of a right triangle, find the remaining side(s) and angle(s) of the triangle

4.9   Law of Sines and Law of Cosines

- Know when it is appropriate to use the Law of Sines, and when it is appropriate to use the Law of Cosines

- Know when the Law of Sines can lead to two possible triangles

- Given three of the six parts of a triangle, find the remaining parts using the Law of Sines

- Given three of the six parts of a triangle, find the remaining parts using the Law of Cosines

4.10   Calculus Preview—The Limit Concept Revisited

- Evaluate the limit of a given function at a given number

Exponential and Logarithmic Functions

5.1 Exponential Functions

- Sketch the graph of an exponential function

- Find the value of b such that the graph of $f(x) = b^x$ passes through a given point

- Determine the range of a given exponential function

- Find the x- and y-intercepts of a given exponential function

- Use a graph to solve an exponential inequality

- Determine if a given exponential function is even or odd or neither

- Find the exact value of the solution of an exponential equation

- Find the x-intercept(s) of the graph of an exponential function

5.2   Logarithmic Functions

- Write a given exponential expression as an equivalent logarithmic expression
- Write a given logarithmic expression as an equivalent exponential expression
- Find the exact value of a given logarithmic expression
- Find the exact value of a given exponential expression involving a logarithm
- Find the value of b such that the graph of $f(x) = \log_b x$ passes through a given point
- Find the domain of a given logarithmic function
- Find the x-intercept of a given logarithmic function
- Find the asymptote of a given logarithmic function
- Use a graph to solve a given inequality involving logarithms
- Use the laws of logarithms to rewrite a given logarithmic expression as a single logarithm
- Use the laws of logarithms to write a given expression involving products, quotients and powers as an expression that contains no products, quotients, or powers
- Find the exact solution of an equation that involves logarithms
- Use the natural logarithm to solve an equation in which the unknown is in the exponent

5.3   Exponential and Logarithmic Models

- Solve word problems involving population growth
- Solve word problems involving radioactive decay and half-life
- Solve word problems involving carbon dating
- Solve word problems involving Newton's law of cooling/warming
- Solve word problems involving compound interest
- Solve word problems involving the Richter scale
- Solve word problems involving the pH of a chemical solution

5.4   Calculus Preview—The Number e

- Find the derivative of a simple logarithmic function
- Find the derivative of a simple exponential function
- Verify identities involving the hyperbolic functions

Conic Sections

6.1 The Parabola

- Find the vertex, focus, directrix, and axis of the parabola determined by a given equation

- Given the equation of a parabola, sketch its graph

- Given information about the vertex, focus, and directrix of a parabola, find the equation of the parabola

- Find the x- and y-intercepts of a parabola

- Determine the focal width of a given parabola

6.2 The Ellipse

- Find the center, foci, vertices, endpoints of the minor axis, and eccentricity of the ellipse determined by a given equation

- Given the equation of an ellipse, sketch its graph

- Given information about the center, vertices, foci, and endpoints of the minor axis, find the equation of the ellipse

- Determine the focal width of a given ellipse

6.3   The Hyperbola

- Find the center, foci, vertices, asymptotes, and eccentricity of the hyperbola determined by a given equation

- Given the equation of a hyperbola, sketch its graph

- Given information about the center, vertices, foci, and asymptotes, find the equation of the hyperbola

- Determine the focal width of a given hyperbola

6.4   Polar Coordinates

- Plot the point with given polar coordinates

- Find alternative polar coordinates of a point having given polar coordinates

- Find the rectangular coordinates of a point given in polar coordinates

- Find the polar coordinates of a point given in rectangular coordinates

- Find a polar equation that has the same graph as a given rectangular equation

6.5   Graphs of Polar Equations

- Recognize the graphs and equations of a cardioid, a limaçon with an interior loop, a dimpled limaçon, a convex limaçon, a rose curve with an even number of petals, a rose curve with an odd number of petals, a circle centered on a coordinate axis that passes through the origin, and a lemniscate

- Identify by name the graph of a given polar equation

- Find the points of intersection of the graphs of a given pair of polar equations

6.6   Conic Sections in Polar Coordinates

- Given the polar equation of a conic, determine its eccentricity

- Given the polar equation of a conic, sketch its graph

- Given the polar equation of a conic, find the rectangular equation

- Find the polar equation of a conic, given its eccentricity and directrix

- Find the polar equation of a parabola with focus at the origin, given the polar coordinates of its vertex

6.7 │ Calculus Preview—Parametric Equations │

- Sketch the curve that has a given set of parametric equations

- Eliminate the parameter from a given set of parametric equations and obtain a rectangular equation having the same graph

- Compare the graphs of a rectangular equation and a set of parametric equations whose graph lies on the graph of the rectangular equation

- Recognize the role played by the domain of the parameter in the set of parametric equations of a circle centered at the origin

- Find the x- and y-intercepts of the graph determined by a set of parametric equations

PART IV

Selected Solutions

Chapter 1

Inequalities, Equations, and Graphs

1.1 | The Real Line |

3. Since "nonnegative" means positive or 0, the statement is equivalent to $a + b \geq 0$.

6. $c - 1 \leq 5$

9. $[5, \infty)$

12. $(-5, -3]$

15. $-7 \leq x \leq 9$

18. $x \geq -5$

21. Using the properties of inequalities, we have

$$\frac{3}{2}x + 4 \leq 10$$

$$\frac{3}{2}x + 4 - 4 \leq 10 - 4 \qquad \leftarrow \text{ by}(ii)$$

$$\frac{3}{2}x \leq 6$$

$$\left(\frac{2}{3}\right)\frac{3}{2}x \leq \left(\frac{2}{3}\right)6$$

$$x \leq 4.$$

In interval notation this is $(-\infty, 4]$, with graph

24. Using the properties of inequalities, we have

$$-(1 - x) \geq 2x - 1$$
$$-1 + x \geq 2x - 1 \qquad \leftarrow \text{ by}(ii)$$
$$1 - 1 + x \geq 1 + 2x - 1$$
$$x \geq 2x$$
$$-x + x \geq -x + 2x \qquad \leftarrow \text{ by}(ii)$$
$$0 \geq x.$$

To express the solution in interval notation we rewrite the last inequality as $x \leq 0$. In interval notation this is $(-\infty, 0]$ with graph

27. Using the properties of inequalities, we have

$$-\frac{20}{3} < \frac{2}{3}x < 4$$

$$\frac{3}{2}\left(-\frac{20}{3}\right) < \frac{3}{2}\left(\frac{2}{3}x\right) < \frac{3}{2}(4) \qquad \leftarrow \text{ by}(ii)$$

$$-10 < x < 6.$$

In interval notation this is $(-10, 6)$, with graph

30. Using the properties of inequalities, we have

$$3 < x + 4 \leq 10$$

$$3 - 4 < x + 4 - 4 \leq 10 - 4 \qquad \leftarrow \text{ by}(i)$$

$$-1 < x \leq 6.$$

In interval notation this is $(-1, 6]$, with graph

33. Using the properties of inequalities, we have

$$-1 \leq \frac{x - 4}{4} < \frac{1}{2}$$

$$4(-1) \leq 4\left(\frac{x - 4}{4}\right) < 4\left(\frac{1}{2}\right) \qquad \leftarrow \text{ by}(ii)$$

$$-4 \leq x - 4 < 2$$

$$-4 + 4 \leq x - 4 + 4 < 2 + 4 \qquad \leftarrow \text{ by}(i)$$

$$0 \leq x < 6.$$

In interval notation this is $[0, 6)$, with graph

36. First, rewrite the inequality with all nonzero terms to the left of the inequality symbol and then factor.

$$x^2 - 16 \geq 0$$
$$(x + 4)(x - 4) \geq 0.$$

Placing $x = -4$ and $x = 4$ on the number line determines three intervals. The sign chart below, in turn, determines the graph of the solution set.

$x + 4$	$-$	0	$+$	$+$	$+$
$x - 4$	$-$	$-$	$-$	0	$+$
$(x + 4)(x - 4)$	$+$	0	$-$	0	$+$

The solution set is $(-\infty, -4] \cup [4, \infty)$.

39. Factoring, we have

$$x^2 - 8x + 12 < 0$$
$$(x - 2)(x - 6) < 0.$$

Placing $x = 2$ and $x = 6$ on the number line determines three intervals. The sign chart below, in turn, determines the graph of the solution set.

$x - 2$	$-$	0	$+$	$+$	$+$
$x - 6$	$-$	$-$	$-$	0	$+$
$(x - 2)(x - 6)$	$+$	0	$-$	0	$+$

The solution set is $(2, 6)$.

42. First, rewrite the inequality with all nonzero terms to the left of the inequality symbol and then factor.

$$4x^2 > 9x + 9$$
$$4x^2 - 9x - 9 > 0$$
$$(4x + 3)(x - 3) > 0.$$

Placing $x = -\frac{3}{4}$ and $x = 3$ on the number line determines three intervals. The following sign chart, in turn, determines the graph of the solution set.

$$
\begin{array}{lccccc}
4x + 3 & - & 0 & + & + & + \\
x - 3 & - & - & - & 0 & + \\
(4x + 3)(x - 3) & + & 0 & - & 0 & +
\end{array}
$$

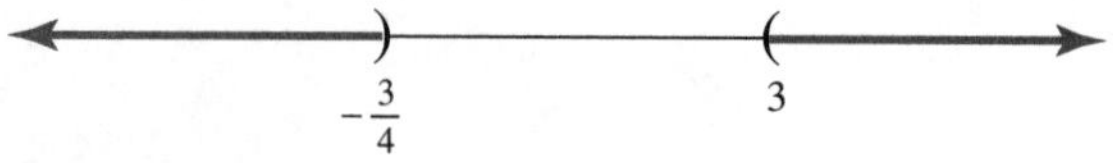

The solution set is $(-\infty, -\frac{3}{4}) \cup (3, \infty)$.

45. Factoring, we have

$$(x^2 - 1)(x^2 - 4) \le 0$$
$$(x + 1)(x - 1)(x + 2)(x - 2) \le 0.$$

Placing $x = -2$, $x = -1$, $x = 1$, and $x = 2$ on the number line determines five intervals. The sign chart below, in turn, determines the graph of the solution set.

$$
\begin{array}{lcccccccccc}
x + 2 & - & 0 & + & + & & + & & + & + & + & + \\
x + 1 & - & - & - & 0 & & + & & + & + & + & + \\
x - 1 & - & - & - & - & & - & & 0 & + & + & + \\
x - 2 & - & - & - & - & & - & & - & - & 0 & + \\
(x + 2)(x + 1)(x - 1)(x - 2) & + & 0 & - & 0 & & + & & 0 & - & 0 & +
\end{array}
$$

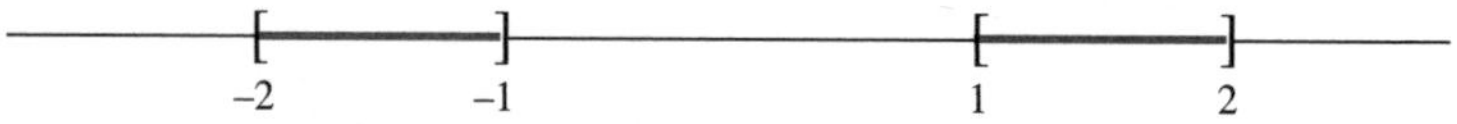

The solution set is $[-2, -1] \cup [1, 2]$.

48. Since 10 is positive and $x^2 + 2$ is positive for all values of x, $10/(x^2 + 2) > 0$ for all real x. The solution set is $(-\infty, \infty)$ and the graph is

51. First, rewrite the left-hand side of the inequality as a simple fraction.

$$\frac{x + 1}{x - 1} + 2 > 0$$
$$\frac{x + 1 + 2(x - 1)}{x - 1} > 0$$
$$\frac{3x - 1}{x - 1} > 0.$$

Placing $x = \frac{1}{3}$ and $x = 1$ on the number line determines three intervals. The sign chart below, in turn, determines the graph of the solution set.

$$
\begin{array}{lcccc}
x - 1 & - & 0 & + & + & + \\
x + 3 & - & - & - & 0 & + \\
(3x - 1)/(x - 1) & + & 0 & - & \text{undefined} & +
\end{array}
$$

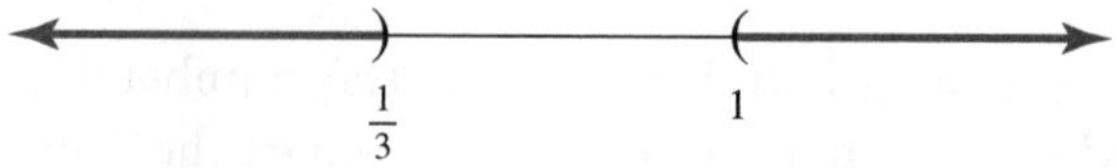

The solution set is $(-\infty, \frac{1}{3}) \cup (1, \infty)$.

54. The left-hand side is already in the proper form. Placing $x = -1$, $x = 0$, and $x = 1$ on the number line determines four intervals. The sign chart below, in turn, determines the graph of the solution set.

$1 + x$	$-$	0	$+$	$+$	$+$	$+$	$+$
$1 - x$	$+$	$+$	$+$	$+$	$+$	0	$-$
x	$-$	$-$	$-$	0	$+$	$+$	$+$
$((1+x)(1-x)/x)$	$+$	0	$-$	undefined	$+$	0	$-$

The solution set is $[-1, 0) \cup [1, \infty)$.

57. First, rewrite the left-hand side of the inequality as a simple fraction.

$$\frac{2}{x+3} + -\frac{1}{x+1} < 0$$
$$\frac{2(x+1) - (x+3)}{(x+3)(x+1)} > 0$$
$$\frac{x-1}{(x+3)(x+1)} > 0.$$

Placing $x = -3$, $x = -1$, and $x = 1$ on the number line determines four intervals. The sign chart below, in turn, determines the graph of the solution set.

$x - 1$	$-$	$-$	$-$	$-$	$-$	0	$+$
$x + 3$	$-$	0	$+$	$+$	$+$	$+$	$+$
$x + 1$	$-$	$-$	$-$	0	$+$	$+$	$+$
$((x-1)/(x+3)(x+1))$	$-$	undefined	$+$	undefined	$-$	0	$+$

The solution set is $(-\infty, -3) \cup (-1, 1)$.

60. The sides of the resulting rectangle are $x + 2$ and $x + 5$. An implicit condition in this case is $x > 0$. (*See the section on Implicit Conditions in a Word Problem in Part I,* **Algebra Topics**, *of this manual.*) We want $(x + 5)(x + 2) < 130$. To solve this inequality, we move all nonzero terms to the left-hand side of the inequality, expand the left-hand side, and factor the result.

$$(x + 5)(x + 2) < 130$$
$$x^2 + 7x + 10 - 130 < 0$$
$$x^2 + 7x - 120 < 0$$
$$(x + 15)(x - 8) < 0$$

Since $x > 0$, we can ignore $x = -15$ and work with a number line that extends from 0 to ∞ and contains 8.

$$
\begin{array}{c|ccc}
x + 15 & + & + & + \\
x - 8 & - & 0 & + \\
(x + 15)(x - 8) & - & 0 & +
\end{array}
$$

The possible lengths of the original square are between 0 and 8 inches, with neither 0 nor 8 included.

63. Let x be the width of the flower bed. Then its length is $2x$ and the area is $(2x)x = 2x^2$. An implicit condition in this case is $x > 0$. (*See the note about implicit conditions in the* **Algebra Topics** *part of this manual.*) We want $2x^2 > 98$. Rewrite the inequality with all nonzero terms on the left-hand side of the inequality symbol and then factor.

$$
\begin{aligned}
2x^2 &> 98 \\
2x^2 - 98 &> 0 \\
2(x + 7)(x - 7) &> 0
\end{aligned}
$$

Since $x > 0$, we can ignore $x = -7$ and work with a number line that extends from 0 to ∞ and contains 7.

$$
\begin{array}{c|ccc}
x + 7 & + & + & + \\
x - 7 & - & 0 & + \\
2(x + 7)(x - 7) & - & 0 & +
\end{array}
$$

The width of the flower bed must be greater than 7 m.

66. Letting $g = 32$, $v_0 = 72$, and $s_0 = 0$ we want to solve $-\frac{1}{2}(32)t^2 + 72t > 80$. An implicit condition in this case is $t \geq 0$. (*See the note about implicit conditions in the* **Algebra Topics** *part of this manual.*) Rewrite the inequality with all nonzero terms on the left-hand side of the inequality symbol and then factor.

$$
\begin{aligned}
-\frac{1}{2}(32)t^2 + 72t &> 80 \\
-16t^2 + 72t - 80 &> 0 \\
16t^2 - 72t + 80 &< 0 \qquad \leftarrow \text{by } (iii) \\
8(2t^2 - 9t + 10) &< 0 \\
8(t - 2)(2t - 5) &< 0
\end{aligned}
$$

Using $t \geq 0$ we place $t = 0$, $t = 2$, and $t = \frac{5}{2}$ on the number line extending to the right from 0. The sign chart below is used to determine the solution set.

$$
\begin{array}{c|ccccccc}
t - 2 & - & & - & & 0 & + & + & + \\
2t - 5 & - & & - & & - & - & 0 & + \\
8(t - 2)(2t - 5) & + & & + & & 0 & - & 0 & +
\end{array}
$$

The rocket will be more that 80 feet above the ground between $t = 2$ and $t = \frac{5}{2}$ seconds.

1.2   Absolute Value

3. Since $\sqrt{63} < \sqrt{64} = 8$, $8 - \sqrt{63}$ is positive and so $|8 - \sqrt{63}| = 8 - \sqrt{63}$.

6. $\big||-3| - |10|\big| = |3 - 10| = |-7| = 7$

9. If $x < 6$, then $x - 6 < 0$ and $|x - 6| = -(x - 6) = 6 - x$.

12. Using the Properties of Absolute Value in the text we have for $x \neq y$

$$\begin{aligned}
\left|\frac{x - y}{y - x}\right| &= \frac{|x - y|}{|y - x|} \qquad \leftarrow \text{ by}(iv) \\
&= \frac{|x - y|}{|-(x - y)|} \\
&= \frac{|x - y|}{|x - y|} \qquad \leftarrow \text{ by}(i) \\
&= 1.
\end{aligned}$$

15. Since x is in $(3, 4]$, $3 < x \leq 4$, so $x - 2 > 0$ and $x - 5 < 0$. Thus,

$$|x - 2| + |x - 5| = x - 2 - (x - 5) = x - 2 - x + 5 = 3.$$

18. Since x is in $(0, 1)$, $0 < x < 1$, so $x + 1 > 0$ and $x - 3 < 0$. Thus,

$$|x + 1| - |x - 3| = x + 1 - [-(x - 3)] = x + 1 + x - 3 = 2x - 2.$$

21. $d(3, 7) = |7 - 3| = |4| = 4$

$$m = \frac{3 + 7}{2} = \frac{10}{2} = 5$$

24. $d\left(-\dfrac{1}{4}, \dfrac{7}{4}\right) = \left|\dfrac{7}{4} - \left(-\dfrac{1}{4}\right)\right| = \left|\dfrac{7}{4} + \dfrac{1}{4}\right| = \left|\dfrac{8}{4}\right| = |2| = 2$

$$m = \frac{-1/4 + 7/4}{2} = \frac{6/4}{2} = \frac{6}{4(2)} = \frac{6}{8} = \frac{3}{4}$$

27. Since $a = 4$ and $d(a, m) = \pi$, we have $m - a = \pi$ or $m = 4 + \pi$. Then

$$m = \frac{a + b}{2} = \frac{4 + b}{2} = 2 + \frac{1}{2}b = 4 + \pi,$$

so $\frac{1}{2}b = 2 + \pi$ and $b = 4 + 2\pi$.

30. The given equation is equivalent to

$$5v - 4 = 7 \qquad \text{or} \qquad 5v - 4 = -7.$$

From $5v - 4 = 7$ we obtain $5v = 7 + 4 = 11$, so $v = \frac{11}{5}$. From $5v - 4 = -7$ we obtain $5v = -7 + 4 = -3$, so $v = -\frac{3}{5}$. Therefore, the solutions are $\frac{11}{5}$ and $-\frac{3}{5}$.

33. The given equation is equivalent to

$$\frac{x}{x-1} = 2 \qquad \text{or} \qquad \frac{x}{x-1} = -2.$$

From $x/(x-1) = 2$ we obtain $x = 2(x-1) = 2x-2$, so $x = 2$. From $x/(x-1) = -2$ we obtain $x = -2(x-1) = -2x+2$, so $3x = 2$ and $x = \frac{2}{3}$. Therefore, the solutions are 2 and $\frac{2}{3}$.

36. The inequality $|3x| > 18$ is equivalent to

$$3x > 18 \qquad \text{or} \qquad 3x < -18.$$

Solving $3x > 18$ we obtain $x > 18/3 = 6$. In interval notation this is $(6, \infty)$. Solving $3x < -18$ we obtain $x < -18/3 = -6$. In interval notation this is $(-\infty, -6)$. The solution set is $(-\infty, -6) \cup (6, \infty)$. The graph of the solution set is

39. The inequality $|2x - 7| \leq 1$ is equivalent to $-1 \leq 2x - 7 \leq 1$. Solving, we obtain

$$-1 + 7 \leq 2x - 7 + 7 \leq 1 + 7$$
$$6 \leq 2x \leq 8$$
$$\tfrac{1}{2}(6) \leq \tfrac{1}{2}(2x) \leq \tfrac{1}{2}(8)$$
$$3 \leq x \leq 4.$$

The solution set is $[3, 4]$ and the graph of the solution set is

42. The inequality $|6x + 4| > 4$ is equivalent to

$$6x + 4 > 4 \qquad \text{or} \qquad 6x + 4 < -4.$$

Solving $6x + 4 > 4$ we obtain $6x > 0$, so $x > 0$. In interval notation this is $(0, \infty)$. Solving $6x + 4 < -4$ we obtain $6x < -8$, so $x < -\frac{8}{6} = -\frac{4}{3}$. In interval notation this is $(-\infty, -\frac{4}{3})$. The solution set is $(-\infty, -\frac{4}{3}) \cup (0, \infty)$. The graph of the solution set is

45. The inequality is equivalent to $-0.01 < x - 5 < 0.01$. Solving, we obtain

$$-0.01 + 5 < x - 5 + 5 < 0.01 + 5$$
$$4.99 < x < 5.01.$$

The solution set is $(4.99, 5.01)$ and the graph of the solution set is

48. Since the solution set is a finite open interval, the form of the inequality will be $|x - b| < a$, where b is the midpoint, m, of the interval and a is the distance from m to b. Thus,

$$m = \frac{1 + 2}{2} = \frac{3}{2}$$

and

$$d(m, b) = b - \frac{3}{2} = \frac{1}{2}d(1, 2) = \frac{1}{2}(2 - 1) = \frac{1}{2}.$$

Then $b = \frac{1}{2} + \frac{3}{2} = \frac{4}{2} = 2$ and the inequality is $\left|x - \frac{3}{2}\right| < \frac{1}{2}$.

51. We want all numbers x such that the distance from x to -3 is greater than 2. That is, we want $d(-3, x) > 2$ or $|x - (-3)| \geq 2$. In interval notation this is $(-\infty, -5] \cup [-1, \infty)$.

54. If your midterm score is M, then the average of M and 72 is $\frac{1}{2}(M + 72)$. Thus, you want

$$80 \leq \tfrac{1}{2}(M + 72) \leq 89$$

$$80 \leq \tfrac{1}{2}M + 36 \leq 89$$

$$80 - 36 \leq \tfrac{1}{2}M \leq 89 - 36$$

$$44 \leq \tfrac{1}{2}M \leq 53$$

$$2(44) \leq 2(\tfrac{1}{2}M) \leq 2(53)$$

$$88 \leq M \leq 106.$$

Assuming that the highest score you can get on a test is 100 (100%), you will need to get a score from 88% to 100%, inclusive, on the midterm to have a mid-semester grade of B.

1.3   The Rectangular Coordinate System

3.

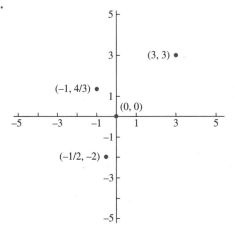

6. Since (a, b) is in quadrant I, $a > 0$ and $b > 0$. Thus, $-b < 0$, and by Figure 1.3.3 in the text, $(a, -b)$ is in quadrant IV.

9. Since (a, b) is in quadrant I, $a > 0$ and $b > 0$. Thus, $-b < 0$, and by Figure 1.3.3 in the text, $(-b, a)$ is in quadrant II.

12. Since (a, b) is in quadrant I, $a > 0$ and $b > 0$. Thus, $-b < 0$, and by Figure 1.3.3 in the text, $(b, -b)$ is in quadrant IV.

15. Since (a, b) is in quadrant I, $a > 0$ and $b > 0$. Thus, $-a < 0$, and by Figure 1.3.3 in the text, $(b, -a)$ is in quadrant IV.

18. $A : (-5, 0)$; $B : (-3, -1)$; $C : (0, -3)$; $D : (-2, -2)$; $E : (3, 3)$; $F : (3, \frac{1}{2})$; $G : (6, -1)$

21. Since $xy = 0$ when $x = 0$ or when $y = 0$, the points that satisfy the condition constitute the coordinate axes.

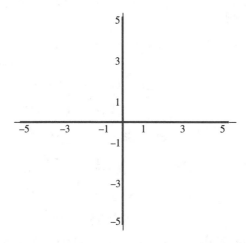

24. Since $x \leq 2$ and $y \geq -1$, the region lies to the left of the line $x = 2$ and above the line $y = -1$. The lines $x = 2$ and $y = -1$ are included, and are thus shown as solid lines in the figure.

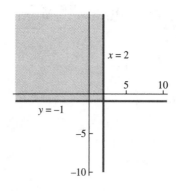

27. $d(A, B) = \sqrt{(-3 - 1)^2 + (4 - 2)^2} = \sqrt{(-4)^2 + 2^2} = \sqrt{16 + 4} = \sqrt{20} = 2\sqrt{5}$

30. $d(A, B) = \sqrt{(-5 - (-12))^2 + (-7 - (-3))^2} = \sqrt{(-5 + 12)^2 + (-7 + 3)^2}$
$$= \sqrt{7^2 + (-4)^2} = \sqrt{49 + 16} = \sqrt{65}$$

33. Since a right triangle satisfies the Pythagorean Theorem, we compute

$$d(A, B) = \sqrt{(-3-8)^2 + (-1-1)^2} = \sqrt{(-11)^2 + (-2)^2} = \sqrt{121+4} = \sqrt{125},$$

$$d(B, C) = \sqrt{(10-(-3))^2 + (5-(-1))^2} = \sqrt{13^2 + 6^2} = \sqrt{169+36} = \sqrt{205},$$

and

$$d(C, A) = \sqrt{(8-10)^2 + (1-5)^2} = \sqrt{(-2)^2 + (-4)^2} = \sqrt{4+16} = \sqrt{20}.$$

The sum of the squares of the shorter sides is $(\sqrt{125})^2 + (\sqrt{20})^2 = 125 + 20 = 145$, which is not $(\sqrt{205})^2 = 205$, the square of the third side. Thus, the three points do not satisfy the Pythagorean Theorem and the triangle is not a right triangle.

36. Since a right triangle satisfies the Pythagorean Theorem, we compute

$$d(A, B) = \sqrt{(1-4)^2 + (1-0)^2} = \sqrt{(-3)^2 + 1^2} = \sqrt{9+1} = \sqrt{10},$$

$$d(B, C) = \sqrt{(2-1)^2 + (3-1)^2} = \sqrt{1^2 + 2^2} = \sqrt{1+4} = \sqrt{5},$$

and

$$d(C, A) = \sqrt{(4-2)^2 + (0-3)^2} = \sqrt{2^2 + (-3)^2} = \sqrt{4+9} = \sqrt{13}.$$

The sum of the squares of the shorter sides is $(\sqrt{10})^2 + (\sqrt{5})^2 = 10 + 5 = 15$, which is not $(\sqrt{13})^2 = 13$, the square of the third side. Thus, the three points do not satisfy the Pythagorean Theorem and the triangle is not a right triangle.

39. (a) If P is equidistant from A and B, then $d(A, P) = d(B, P)$ or

$$\sqrt{(x-(-1))^2 + (y-2)^2} = \sqrt{(x-3)^2 + (y-4)^2}$$

Squaring both side and simplifying we have

$$(x+1)^2 + (y-2)^2 = (x-3)^2 + (y-4)^2$$
$$x^2 + 2x + 1 + y^2 - 4y + 4 = x^2 - 6x + 9 + y^2 - 8y + 16$$
$$2x + 1 - 4y + 4 = -6x + 9 - 8y + 16$$
$$2x - 4y + 5 = -6x - 8y + 25$$
$$8x + 4y = 20$$
$$2x + y = 5.$$

Note: Be careful when squaring both sides of an equation—you may introduce *extraneous solutions*. (See **Extraneous Solutions** in the Algebra Topics part of this manual.) In this case, no extraneous solutions were introduced because the expressions under the square root signs were both nonnegative.

(b) The set of points in the plane that are equidistant from two given distinct points is the perpendicular bisector of the line segment joining A and B.

42. Let A be the point $(1/\sqrt{2}, 1/\sqrt{2})$, B be the point $(0.25, 0.97)$, and O be the origin. Then

$$d(O, A) = \sqrt{(1/\sqrt{2} - 0)^2 + (1/\sqrt{2} - 0)^2} = \sqrt{\tfrac{1}{2} + \tfrac{1}{2}} = \sqrt{1} = 1$$

$$d(O, B) = \sqrt{(0.25 - 0)^2 + (0.97 - 0)^2} = \sqrt{0.0625 + 0.9409} = \sqrt{1.0034} \approx 1.0017.$$

We see that $(1/\sqrt{2}, 1/\sqrt{2})$ is closer to the origin than $(0.25, 0.97)$.

45. Using the midpoint formula, we find that the coordinates of the midpoint are given by

$$\left(\frac{-1-8}{2}, \frac{0+5}{2} \right) \quad \text{or} \quad \left(-\frac{9}{2}, \frac{5}{2} \right).$$

48. Using the midpoint formula, we find that the coordinates of the midpoint are given by

$$\left(\frac{x + (-x)}{2}, \frac{x + (x+2)}{2} \right) \quad \text{or} \quad \left(0, \frac{2x+2}{2} \right) \quad \text{or} \quad (0, x+1).$$

51. If the coordinates of B are (x, y) then, by the midpoint formula,

$$-1 = \frac{1}{2}(5 + x) \quad \text{or} \quad x = -2 - 5 = -7$$

and

$$-1 = \frac{1}{2}(8 + y) \quad \text{or} \quad y = -2 - 8 = -10.$$

Thus, B is $(-7, -10)$.

54. Using the midpoint formula, we find that the coordinates of the midpoint are given by

$$\left(\frac{5 + (-5)}{2}, \frac{2 + (-6)}{2} \right) \quad \text{or} \quad (0, -2).$$

Since any point on the x-axis has y-coordinate 0, let $(x, 0)$ be a point on the x-axis that is 3 units from $(0, -2)$. By the distance formula,

$$d\big((0, -2), (x, 0)\big) = \sqrt{(x - 0)^2 + (0 - (-2))^2} = \sqrt{x^2 + 4} = 3.$$

Squaring both sides, we have $x^2 + 4 = 9$, $x^2 = 5$, and $x = \pm\sqrt{5}$. The points on the x-axis are $(-\sqrt{5}, 0)$, and $(\sqrt{5}, 0)$.

57. We first let $P_2(x_2, y_2)$ be the midpoint of the line segment joining $A(3, 6)$ and $B(5, 8)$. Then, by the midpoint formula, $x_2 = \frac{1}{2}(3 + 5) = \frac{1}{2}(8) = 4$ and $y_2 = \frac{1}{2}(6 + 8) = \frac{1}{2}(14) = 7$. Thus, P_2 is $(4, 7)$. Now, P_1 is the midpoint of $A(3, 6)$ and $P_2(4, 7)$, so $x_1 = \frac{1}{2}(3 + 4) = \frac{7}{2}$ and $y_1 = \frac{1}{2}(6 + 7) = \frac{13}{2}$. Thus, P_1 is $\left(\frac{7}{2}, \frac{13}{2} \right)$. Finally, $P_3(x_3, y_3)$ is the midpoint of $P_2(4, 7)$ and $B(5, 8)$, so $x_3 = \frac{1}{2}(4 + 5) = \frac{9}{2}$ and $y_3 = \frac{1}{2}(7 + 8) = \frac{15}{2}$. Thus, P_3 is $\left(\frac{9}{2}, \frac{15}{2} \right)$.

1.4   Circles and Graphs

3. Writing the equation in the form

$$x^2 + (y-3)^2 = 7^2$$

we see that the center is $(0, 3)$ and the radius is 7.

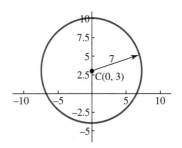

6. Writing the equation in the form

$$(x-(-3))^2 + (y-5)^2 = 5^2$$

we see that the center is $(-3, 5)$ and the radius is 5.

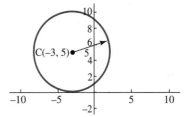

9. In order to find the center and radius, we want to write the equation in standard form. To do this, we rearrange the terms and complete the square:

$$x^2 + y^2 + 2x - 4y - 4 = 0$$
$$[x^2 + 2x \qquad] + [y^2 - 4y \qquad] = 4$$
$$[x^2 + 2x + (\tfrac{2}{2})^2] + [y^2 - 4y + (\tfrac{-4}{2})^2] = 4 + (\tfrac{2}{2})^2 + (\tfrac{-4}{2})^2$$
$$(x^2 + 2x + 1) + (y^2 - 4y + 4) = 4 + 1 + 4 = 9$$
$$(x+1)^2 + (y-2)^2 = 3^2.$$

From the last equation, we see that the circle is centered at $(-1, 2)$ and has radius 3.

12. In order to find the center and radius, we want to write the equation in standard form. To do this, we rearrange the terms and complete the square:

$$x^2 + y^2 + 3x - 16y + 63 = 0$$
$$[x^2 + 3x \qquad] + [y^2 - 16y \qquad] = -63$$
$$[x^2 + 3x + (\tfrac{3}{2})^2] + [y^2 - 16y + (\tfrac{-16}{2})^2] = -63 + (\tfrac{3}{2})^2 + (\tfrac{-16}{2})^2$$
$$(x^2 + 3x + \tfrac{9}{4}) + (y^2 - 16y + 64) = -63 + \tfrac{9}{4} + 64 = \tfrac{13}{4}$$
$$(x + \tfrac{3}{2})^2 + (y-8)^2 = (\sqrt{13}/2)^2.$$

From the last equation, we see that the circle is centered at $(-\tfrac{3}{2}, 8)$ and has radius $\sqrt{13}/2$.

15. Substituting $h = 0$, $k = 0$, and $r = 1$ in equation (2) in the text, we obtain

$$(x-0)^2 + (y-0)^2 = 1^2 \qquad \text{or} \qquad x^2 + y^2 = 1.$$

18. Substituting $h = -9$, $k = -4$, and $r = \tfrac{3}{2}$ in equation (2) in the text, we obtain

$$\left(x - (-9)\right)^2 + \left(y - (-4)\right)^2 = \left(\frac{3}{2}\right)^2 \qquad \text{or} \qquad (x+9)^2 + (y+4)^2 = \frac{9}{4}.$$

21. We identify $h = 0$ and $k = 0$. Since the radius is the distance from the center to a point on the circle, we have

$$\sqrt{(-1-0)^2 + (-2-0)^2} = \sqrt{1+4} = \sqrt{5}.$$

Thus, the equation of the circle is

$$\left(x - (-1)\right)^2 + \left(y - (-2)\right)^2 = \left(\sqrt{5}\right)^2 \qquad \text{or} \qquad (x+1)^2 + (y+2)^2 = 5.$$

24. Since the graph is tangent to the y-axis, the radius of the circle is 4, as seen in the figure. Thus, the equation of the circle is

$$\left(x - (-4)\right)^2 + \left(y - 3\right)^2 = 4^2$$

or

$$(x+4)^2 + (y-3)^2 = 16.$$

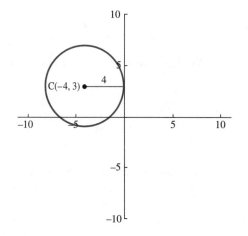

27. Squaring both sides of the equation and writing in the standard form of the equation of a circle, we get

$$x^2 = 1 - (y-1)^2 \quad \text{or} \quad x^2 + (y-1)^2 = 1 = 1^2.$$

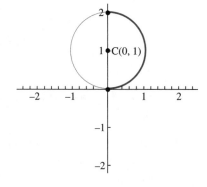

The graph of this equation is a circle with center $(0, 1)$ and radius 1. Since the semicircle is defined by $x = \sqrt{1 - (y-1)^2}$, we see that $x \geq 0$. The graph is thus the right half of the circle. The figure shows the circle in light blue and the semicircle as a darker and thicker curve.

30. To find the lower half of the circle, we solve for y and use the negative square root of the result:

$$(y-1)^2 = 9 - (x-5)^2$$
$$y - 1 = -\sqrt{9 - (x-5)^2}$$
$$y = 1 - \sqrt{9 - (x-5)^2}.$$

To find the left half of the circle, we solve for x and use the negative square root of the result:

$$(x-5)^2 = 9 - (y-1)^2$$
$$x - 5 = -\sqrt{9 - (y-1)^2}$$
$$x = 5 - \sqrt{9 - (y-1)^2}.$$

33. The inequality represents the region between the circles $x^2 + y^2 = 1$ and $x^2 + y^2 = 4$. Because the inequalities both involve "$\leq$"(as opposed to "$<$"), the two circles are part of the region.

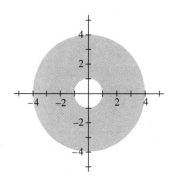

36. To find the x-intercepts, we set $y = 0$ in the equation. This gives $x^2 + 5x = 0$ or $x(x + 5) - 0$. Thus, $x = 0$ and $x = -5$, so the x-intercepts are $(0, 0)$ and $(-5, 0)$. To find the y-intercepts we set $x = 0$ in the equation. This gives $y^2 - 6y = 0$ or $y(y - 6) = 0$. Thus, $y = 0$ and $y = 6$, so the y-intercepts are $(0, 0)$ and $(0, 6)$. [Note that whenever the origin is an intercept, it is both an x- and a y-intercept.]

39. *Intercepts*: Setting $y = 0$ in $-x + 2y = 1$ we get $-x = 1$ or $x = -1$, so the x-intercept is $(-1, 0)$. Setting $x = 0$ we get $2y = 1$ or $y = \frac{1}{2}$, so the y-intercept is $(0, \frac{1}{2})$.

Symmetry with respect to the x-axis: Replacing y by $-y$ in the equation $-x + 2y = 1$ we get $-x + 2(-y) = 1$ or $-x - 2y = 1$. This is not equivalent to the original equation, so the graph is not symmetric with respect to the x-axis.

Symmetry with respect to the y-axis: Replacing x by $-x$ in the equation $-x + 2y = 1$ we get $-(-x) + 2y = 1$ or $x + 2y = 1$. This is not equivalent to the original equation, so the graph is not symmetric with respect to the y-axis.

Symmetry with respect to the origin: Replacing x by $-x$ and y by $-y$ in the equation $-x + 2y = 1$ we get $-(-x) + 2(-y) = 1$ or $x - 2y = 1$. This is not equivalent to the original equation, so the graph is not symmetric with respect to the origin.

42. *Intercepts*: Setting $y = 0$ in $y = x^3$ we get $0 = x^3$ or $x = 0$, so the x-intercept is $(0, 0)$. This is also a y-intercept. Setting $x = 0$ we get $y = 0$, so $(0, 0)$ is the only y-intercept.

Symmetry with respect to the x-axis: Replacing y by $-y$ in the equation $y = x^3$ we get $-y = x^3$ or $y = -x^3$. This is not equivalent to the original equation, so the graph is not symmetric with respect to the x-axis.

Symmetry with respect to the y-axis: Replacing x by $-x$ in the equation $y = x^3$ we get $y = (-x)^3$ or $y = -x^3$. This is not equivalent to the original equation, so the graph is not symmetric with respect to the y-axis.

Symmetry with respect to the origin: Replacing x by $-x$ and y by $-y$ in the equation $y = x^3$ we get $-y = (-x)^3$, $y = -(-x^3)$, so $y = x^3$. This is equivalent to the original equation, so the graph is symmetric with respect to the origin.

45. *Intercepts*: Setting $y = 0$ in $y = x^2 - 2x - 2$ we get $0 = x^2 - 2x - 2$. Using the quadratic formula, we find

$$x = \frac{-(-2) \pm \sqrt{(-2)^2 - 4(1)(-2)}}{2(1)} = \frac{2 \pm \sqrt{4 + 8}}{2} = \frac{2 \pm \sqrt{12}}{2} = \frac{2 \pm 2\sqrt{3}}{2} = 1 \pm \sqrt{3}.$$

The x-intercepts are $(1 - \sqrt{3}, 0)$ and $(1 + \sqrt{3}, 0)$. Setting $x = 0$ we get $y = 0^2 - 2(0) - 2 = -2$, so the y-intercept is $(0, -2)$.

Symmetry with respect to the x-axis: Replacing y by $-y$ in the equation $y = x^2 - 2x - 2$ we get $-y = x^2 - 2x - 2$ or $y = -x^2 + 2x + 2$. This is not equivalent to the original equation, so the graph is not symmetric with respect to the x-axis.

Symmetry with respect to the y-axis: Replacing x by $-x$ in the equation $y = x^2 - 2x - 2$ we get $y = (-x)^2 - 2(-x) - 2$ or $y = x^2 + 2x + 2$. This is not equivalent to the original equation, so the graph is not symmetric with respect to the y-axis.

Symmetry with respect to the origin: Replacing x by $-x$ and y by $-y$ in the equation $y = x^2 - 2x - 2$ we get $-y = (-x)^2 - 2(-x) - 2$ or $y = -x^2 - 2x + 2$. This is not equivalent to the original equation, so the graph is not symmetric with respect to the origin.

48. *Intercepts*: Setting $y = 0$ in $y = (x - 2)^2(x + 2)^2$ we get $0 = (x - 2)^2(x + 2)^2$, so the x-intercepts are $(-2, 0)$ and $(2, 0)$. Setting $x = 0$ we get $y = (0 - 2)^2(0 + 2)^2 = 4(4) = 16$, so the y-intercept is $(0, 16)$.

Symmetry with respect to the x-axis: Replacing y by $-y$ in the equation $y = (x - 2)^2(x + 2)^2$ we get $-y = (x - 2)^2(x + 2)^2$ or $y = -(x - 2)^2(x + 2)^2$. This is not equivalent to the original equation, so the graph is not symmetric with respect to the x-axis.

Symmetry with respect to the y-axis: Replacing x by $-x$ in the equation $y = (x - 2)^2(x + 2)^2$ we get

$$y = (-x - 2)^2(-x + 2)^2 = [-(x + 2)]^2[-(x - 2)]^2 = (x + 2)^2(x - 2)^2.$$

This is equivalent to the original equation, so the graph is symmetric with respect to the y-axis.

Symmetry with respect to the origin: Replacing x by $-x$ and y by $-y$ in the equation $y = (x - 2)^2(x + 2)^2$ we get

$$-y = (-x - 2)^2(-x + 2)^2 = [-(x + 2)]^2[-(x - 2)]^2 = (x + 2)^2(x - 2)^2$$

or $y = -(x - 2)^2(x + 2)^2$. This is not equivalent to the original equation, so the graph is not symmetric with respect to the origin.

51. *Intercepts*: Setting $y = 0$ in $4y^2 - x^2 = 36$ we get $4(0)^2 - x^2 = 36$ or $x^2 = -36$. Since the square of a number must be nonnegative, the equation has no solution and there are no x-intercepts. Setting $x = 0$ we get $4y^2 - 0^2 = 36$, $y^2 = 9$, and $y = \pm 3$, so the y-intercepts are $(0, -3)$ and $(0, 3)$.

Symmetry with respect to the x-axis: Replacing y by $-y$ in the equation $4y^2 - x^2 = 36$ we get $4(-y)^2 - x^2 = 36$ or $4y^2 - x^2 = 36$. This is equivalent to the original equation, so the graph is symmetric with respect to the x-axis.

Symmetry with respect to the y-axis: Replacing x by $-x$ in the equation $4y^2 - x^2 = 36$ we get $4y^2 - (-x)^2 = 36$ or $4y^2 - x^2 = 36$. This is equivalent to the original equation, so the graph is symmetric with respect to the y-axis.

Symmetry with respect to the origin: Replacing x by $-x$ and y by $-y$ in the equation $4y^2 - x^2 = 36$ we get $4(-y)^2 - (-x)^2 = 36$ or $4y^2 - x^2 = 36$. This is equivalent to the original equation, so the graph is symmetric with respect to the origin.

54. *Intercepts*: Setting $y = 0$ in $y = (x^2 - 10)/(x^2 + 10)$ we get $0 = (x^2 - 10)/(x^2 + 10)$ so $x^2 - 10 = 0$ and $x = \pm\sqrt{10}$. The x-intercepts are $(-\sqrt{10}, 0)$ and $(\sqrt{10}, 0)$. Setting $x = 0$ we get $y = (0^2 - 10)/(0^2 + 10) = -10/10 = -1$, so the y-intercept is $(0, -1)$.

Symmetry with respect to the x-axis: Replacing y by $-y$ in the equation $y = (x^2 - 10)/(x^2 + 10)$ we get $-y = (x^2 - 10)/(x^2 + 10)$ or $y = -(x^2 - 10)/(x^2 + 10)$. This is not equivalent to the original equation, so the graph is not symmetric with respect to the x-axis.

Symmetry with respect to the y-axis: Replacing x by $-x$ in the equation $y = (x^2 - 10)/(x^2 + 10)$ we get $y = [(-x)^2 - 10]/[(-x)^2 + 10] = (x^2 - 10)/(x^2 + 10)$. This is equivalent to the original equation, so the graph is symmetric with respect to the y-axis.

Symmetry with respect to the origin: Replacing x by $-x$ and y by $-y$ in the equation $y = (x^2 - 10)/(x^2 + 10)$ we get $-y = [(-x)^2 - 10]/[(-x)^2 + 10] = (x^2 - 10)/(x^2 + 10)$, so $y = -(x^2 - 10)/(x^2 + 10)$. This is not equivalent to the original equation, so the graph is not symmetric with respect to the origin.

57. *Intercepts*: Setting $y = 0$ in $y = \sqrt{x} - 3$ we get $0 = \sqrt{x} - 3$ so $3 = \sqrt{x}$ and $x = 9$. Since we squared both sides of the equation to obtain this solution, we must check that $x = 9$ actually is a solution. It is, so the x-intercept is $(9, 0)$. Setting $x = 0$ we get $y = \sqrt{0} - 3 = -3$, so the y-intercept is $(0, -3)$.

Symmetry with respect to the x-axis: Replacing y by $-y$ in the equation $y = \sqrt{x} - 3$ we get $-y = \sqrt{x} - 3$ or $y = -(\sqrt{x} - 3)$. This is not equivalent to the original equation, so the graph is not symmetric with respect to the x-axis.

Symmetry with respect to the y-axis: Since the equation involves $\sqrt{x}$, x must be nonnegative. Hence, the graph of the equation cannot be symmetric with respect to the y-axis.

Symmetry with respect to the origin: Again, since the equation involves $\sqrt{x}$, x must be nonnegative. Hence, the graph of the equation cannot be symmetric with respect to the origin.

60. *Intercepts*: Setting $y = 0$ in $x = |y| - 4$ we get $x = |0| - 4 = -4$ so the x-intercept is $(-4, 0)$. Setting $x = 0$ we get $0 = |y| - 4$, $|y| = 4$, and $y = \pm 4$. Thus, the y-intercepts are $(0, -4)$ and $(0, 4)$.

Symmetry with respect to the x-axis: Replacing y by $-y$ in the equation $x = |y| - 4$ we get $x = |-y| - 4$ or $x = |y| - 4$. This is equivalent to the original equation, so the graph is symmetric with respect to the x-axis.

Symmetry with respect to the y-axis: Replacing x by $-x$ in the equation $x = |y| - 4$ we get $-x = |y| - 4$, so $x = -(|y| - 4)$. This is not equivalent to the original equation, so the graph is not symmetric with respect to the y-axis.

Symmetry with respect to the origin: Replacing x by $-x$ and y by $-y$ in the equation $x = |y| - 4$ we get $-x = |-y| - 4$, $-x = |y| - 4$, and $x = -(|y| - 4)$. This is not equivalent to the original equation, so the graph is not symmetric with respect to the origin.

63. We see that the graph is symmetric with respect to the x-axis, the y-axis, and the origin.

66. We see that the graph is symmetric with respect to the x-axis, the y-axis, and the origin.

69. Since the graph is to be symmetric with respect to the origin, each point in the first quadrant is reflected through the origin to a corresponding point on the graph in the third quadrant.

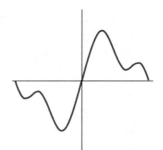

72. Since the graph is to be symmetric with respect to the origin, each point in the second quadrant is reflected through the origin to a corresponding point on the graph in the fourth quadrant, and each point in the third quadrant is reflected through the origin to a corresponding point in the first quadrant.

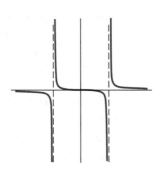

1.5 | Calculus Preview—Algebra and Limits

3. (a) $\dfrac{x^2 - 7x + 6}{x - 1} = \dfrac{(x - 6)(x - 1)}{x - 1} = x - 6$

(b) $\displaystyle\lim_{x \to 1} \dfrac{x^2 - 7x + 6}{x - 1} = \lim_{x \to 1}(x - 6) = 1 - 6 = -5$

6. (a) $\dfrac{x^2 - 8x}{x^2 - 6x - 16} = \dfrac{x(x - 8)}{(x + 2)(x - 8)} = \dfrac{x}{x + 2}$

(b) $\displaystyle\lim_{x \to 8} \dfrac{x^2 - 8x}{x^2 - 6x - 16} = \lim_{x \to 8} \dfrac{x}{x + 2} = \dfrac{8}{8 + 2} = \dfrac{8}{10} = \dfrac{4}{5}$

9. (a) $\dfrac{x^3 - 1}{x^2 + 3x - 4} = \dfrac{(x - 1)(x^2 + x + 1)}{(x - 1)(x + 4)} = \dfrac{x^2 + x + 1}{x + 4}$

(b) $\displaystyle\lim_{x \to 1} \dfrac{x^3 - 1}{x^2 + 3x - 4} = \lim_{x \to 1} \dfrac{x^2 + x + 1}{x + 4} = \dfrac{1^2 + 1 + 1}{5} = \dfrac{3}{5}$

12. (a) $\dfrac{x^4 - 5x^3 + 4x - 20}{x^4 - 5x^3 + x - 5} = \dfrac{x^3(x - 5) + 4(x - 5)}{x^3(x - 5) + (x - 5)} = \dfrac{(x^3 + 4)(x - 5)}{(x^3 + 1)(x - 5)} = \dfrac{x^3 + 4}{x^3 + 1}$

(While $x^3 + 1$ factors with integer coefficients, $x^3 + 4$ does not, so there is no need to further factor the denominator. The fraction $(x^3 + 4)/(x^3 + 1)$ will not further simplify.)

(b) $\displaystyle\lim_{x \to 5} \dfrac{x^4 - 5x^3 + 4x - 20}{x^4 - 5x^3 + x - 5} = \lim_{x \to 5} \dfrac{x^3 + 4}{x^3 + 1} = \dfrac{5^3 + 4}{5^3 + 1} = \dfrac{129}{126} = \dfrac{43}{42}$

15. (a) $\dfrac{(2x+1)^3 - 9}{x-1} = \dfrac{4x^2 + 4x + 1 - 9}{x-1} = \dfrac{4x^2 + 4x - 8}{x-1} = \dfrac{4(x^2 + x - 2)}{x-1}$

$$= \dfrac{4(x+2)(x-1)}{(x-1)} = 4(x+2)$$

(b) $\displaystyle\lim_{x\to1} \dfrac{(2x+1)^3 - 9}{x-1} = \lim_{x\to1} 4(x+2) = 4(1+2) = 12$

18. (a) $\dfrac{(x+1)^3 + (x-1)^3}{x} = \dfrac{(x^3 + 3x^2 + 3x + 1) + (x^3 - 3x^2 + 3x - 1)}{x} = \dfrac{2x^3 + 6x}{x}$

$$= \dfrac{2x(x^2 + 3)}{x} = 2(x^3 + 3)$$

(b) $\displaystyle\lim_{x\to0} \dfrac{(x+1)^3 + (x-1)^3}{x} = \lim_{x\to0} 2(x^3 + 3) = 2(0^3 + 3) = 6$

21. (a) $\dfrac{1}{x-2} - \dfrac{6}{x^2 + 2x - 8} = \dfrac{1}{x-2} - \dfrac{6}{(x-2)(x+4)} = \dfrac{(x+4)-6}{(x-2)(x+4)} = \dfrac{x-2}{(x-2)(x+4)}$

$$= \dfrac{1}{x+4}$$

(b) $\displaystyle\lim_{x\to2} \left[\dfrac{1}{x-2} - \dfrac{6}{x^2 + 2x - 8} \right] = \lim_{x\to2} \dfrac{1}{x+4} = \dfrac{1}{2+4} = \dfrac{1}{6}$

24. (a) $\dfrac{1}{x}\left[\dfrac{1}{9} - \dfrac{1}{x+9} \right] = \dfrac{1}{x}\left[\dfrac{(x+9)-9}{9(x-9)} \right] = \dfrac{x}{x(9)(x-9)} = \dfrac{1}{9(x-9)}$

(b) $\displaystyle\lim_{x\to0} \dfrac{1}{x}\left[\dfrac{1}{9} - \dfrac{1}{x+9} \right] = \lim_{x\to0} \dfrac{1}{9(x-9)} = \dfrac{1}{9(0-9)} = -\dfrac{1}{81}$

27. (a) $\dfrac{\sqrt{x}-3}{x-9} = \dfrac{\sqrt{x}-3}{x-9}\,\dfrac{\sqrt{x}+3}{\sqrt{x}+3} = \dfrac{x-9}{(x-9)(\sqrt{x}+3)} = \dfrac{1}{\sqrt{x}+3}$

(b) $\displaystyle\lim_{x\to9} \dfrac{\sqrt{x}-3}{x-9} = \lim_{x\to9} \dfrac{1}{\sqrt{x}+3} = \dfrac{1}{\sqrt{9}+3} = \dfrac{1}{3+3} = \dfrac{1}{6}$

30. (a) $\dfrac{\sqrt{u+4}-3}{u-5} = \dfrac{\sqrt{u+4}-3}{u-5}\,\dfrac{\sqrt{u+4}+3}{\sqrt{u+4}+3} = \dfrac{u+4-9}{(u-5)(\sqrt{u+4}+3)}$

$$= \dfrac{u-5}{(u-5)(\sqrt{u+4}+3)} = \dfrac{1}{\sqrt{u+4}+3}$$

(b) $\displaystyle\lim_{u\to5} \dfrac{\sqrt{u+4}-3}{u-5} = \lim_{u\to5} \dfrac{1}{\sqrt{u+4}+3} = \dfrac{1}{\sqrt{5+4}+3} = \dfrac{1}{\sqrt{9}+3} = \dfrac{1}{3+3} = \dfrac{1}{6}$

33. (a) $\dfrac{4y^2}{\sqrt{y^2+y+1}-\sqrt{y+1}} = \dfrac{4y^2}{\sqrt{y^2+y+1}-\sqrt{y+1}}\,\dfrac{\sqrt{y^2+y+1}+\sqrt{y+1}}{\sqrt{y^2+y+1}+\sqrt{y+1}}$

$$= \dfrac{4y^2(\sqrt{y^2+y+1}+\sqrt{y+1})}{y^2+y+1-(y+1)} = \dfrac{4y^2(\sqrt{y^2+y+1}+\sqrt{y+1})}{y^2}$$

$$= 4(\sqrt{y^2+y+1}+\sqrt{y+1})$$

(b) $\displaystyle \lim_{y \to 0} \frac{4y^2}{\sqrt{y^2 + y + 1} - \sqrt{y + 1}}$

$$= \lim_{y \to 0} 4(\sqrt{y^2 + y + 1} + \sqrt{y + 1}) = 4(\sqrt{0^2 + 0 + 1} + \sqrt{0 + 1})$$

$$= 4(\sqrt{1} + \sqrt{1}) = 8$$

36. (a) $\displaystyle \frac{\dfrac{3}{(x+1)^2} - \dfrac{3}{(a+1)^2}}{x - a} = \frac{\dfrac{3(a+1)^2 - 3(x+1)^2}{(x+1)^2(a+1)^2}}{x - a} = \frac{3(a^2 + 2a + 1) - 3(x^2 + 2x + 1)}{(x-a)(x+1)^2(a+1)^2}$

$$= \frac{3a^2 + 6a - 3x^2 - 6x}{(x-a)(x+1)^2(a+1)^2} = \frac{(3a^2 - 3x^2) + 6(a - x)}{(x-a)(x+1)^2(a+1)^2}$$

$$= \frac{3(a - x)(a + x) + 6(a - x)}{(x-a)(x+1)^2(a+1)^2} = \frac{3(a - x)[(a+x) + 2]}{(x-a)(x+1)^2(a+1)^2}$$

$$= \frac{-3(x - a)(a + x + 2)}{(x-a)(x+1)^2(a+1)^2} = \frac{-3(a + x + 2)}{(x+1)^2(a+1)^2}$$

39. (a) $\displaystyle \frac{2x(-4x+6)^{1/2} - x^2(\frac{1}{2})(-4x+6)^{-1/2}(-4)}{[(-4x+6)^{1/2}]^2} = \frac{2x(-4x+6)^{1/2} + 2x^2(-4x+6)^{-1/2}}{-4x + 6}$

$$= \frac{2x(-4x+6)^{1/2} + 2x^2(-4x+6)^{-1/2}}{-4x + 6} \cdot \frac{(-4x+6)^{1/2}}{(-4x+6)^{1/2}} = \frac{2x(-4x+6) + 2x^2}{(-4x+6)(-4x+6)^{1/2}}$$

$$= \frac{-8x^2 + 12x + 2x^2}{(-4x+6)^{3/2}} = \frac{-6x^2 + 12x}{(-4x+6)^{3/2}} = \frac{-6x(x - 2)}{(-4x+6)^{3/2}} = \frac{6x(2 - x)}{(-4x+6)^{3/2}}$$

42. Multiply out the right-hand side and solve for y':

$$y' = 2(x - y) - 2(x - y)y'$$
$$2(x - y)y' + y' = 2x - 2y$$
$$[2(x - y) + 1]y' = 2x - 2y$$
$$y' = \frac{2x - 2y}{2x - 2y + 1}.$$

45. Multiply both sides of the equation by $(x - y)^2$ and solve for y':

$$(x - y)(1 + y') - (x + y)(1 - y') = (x - y)^2$$
$$(x - y) + (x - y)y' - (x + y) + (x + y)y' = (x - y)^2$$
$$(x - y)y' + (x + y)y' = (x - y)^2 - (x - y) + (x + y)$$
$$(x - y + x + y)y' = x^2 - 2xy + y^2 - x + y + x + y$$
$$2xy' = x^2 - 2xy + y^2 + 2y$$
$$y' = \frac{x^2 - 2xy + y^2 + 2y}{2x}.$$

Chapter 1 Review Exercises

3. If (a, b) is in quadrant IV, then $a > 0$ and $b < 0$. Thus, (b, a) has a negative first coordinate and positive second coordinate, so it is in quadrant II.

6. If a graph is symmetric with respect to the origin, then whenever (x, y) is on the graph, so is $(-x, -y)$. Thus, if $(-1, 6)$ is on a graph which is symmetric with respect to the origin, then $(1, -6)$ is also on the graph.

9. Using the midpoint formula, we find that the coordinates of the midpoint are given by

$$\left(\frac{4 + (-2)}{2}, \frac{-6 + 0}{2} \right) \qquad \text{or} \qquad (1, -3).$$

Using the distance formula, we find that the distance from $(0, 0)$ to this midpoint is

$$d(0, M) = \sqrt{(1 - 0)^2 + (-3 - 0)^2} = \sqrt{1 + 9} = \sqrt{10}.$$

12. We complete the square to obtain the equation of the circle in standard form:

$$x^2 - 16y + y^2 = 0$$
$$(x^2 - 16y \quad) + (y - 0)^2 = 0$$
$$(x^2 - 16y + 8^2) + (y - 0)^2 = 8^2$$
$$(x - 4)^2 + (y - 0)^2 = 64.$$

The circle is centered at $(8, 0)$ and is thus symmetric with respect to the x-axis.

15. Letting $x = -3$ in $x^2 + y^2 = 25$ we have

$$(-3)^2 + y^2 = 25$$
$$9 + y^2 = 25$$
$$y^2 = 25 - 9 = 16,$$

so $y = \pm 4$, The points on the circle are $(-3, -4)$ and $(-3, 4)$.

18. Letting $x = a$ and $y = a + \sqrt{3}$, we have

$$a + \sqrt{3} = 2a \qquad \text{or} \qquad a = \sqrt{3}.$$

21. False, because 0 is nonnegative but not positive.

24. If $a < b$, then adding $-a$ to both sides of the inequality gives us

$$a - a < b - a \qquad \text{or} \qquad 0 < b - a,$$

so $b - a$ is positive and the statement is true.

27. If $a < 0$ (or a is negative), then $-a$ is positive and $a/(-a)$ is negative. The statement is true.

30. Since the absolute value of any real number is nonnegative, $|4x - 6| \geq 0 > -1$, and the statement is true.

33. The y-coordinate of a point in quadrant III is negative, so $(-3, 7)$ cannot be in quadrant III and the statement is false.

36. Substituting $x = 2$ into the equation $x^2 + y^2 - 10x + 22 = 0$ we obtain $2^2 + y^2 - 10(2) + 22 = 0$ or $y^2 + 6 = 0$. Since $y^2 = -6$ has no real solutions, there is no point on the circle with $x = 2$. The statement is true.

39. Replacing x by $-x$ and y by $-y$ in the equation $x^2 y + 4y = x$ we get $(-x)^2(-y) + 4(-y) = -x$ or $-x^2 y - 4y = -x$. This is equivalent to $x^2 y + 4y = x$, so the graph of the equation is symmetric with respect to the origin and the statement is true.

42. Since multiplication of an inequality by a negative number like -1 changes the sense of the inequality,

$$0(-1) > a(-1) > b(-1), \qquad 0 > -a > -b, \qquad \text{and} \qquad -b < -a < 0.$$

45. Adding 20 to both sides of the inequality, we have

$$x - 10 + 20 > 5 + 20, \qquad \text{so} \qquad x + 10 > 25.$$

48. We solve the inequality for x:

$$3x - 6 \leq 4x - 4$$
$$-3x + (3x - 6) + 4 \leq -3x + (4x - 4) + 4$$
$$-2 \leq x.$$

Thus, $x \geq -2$.

51. If $d(a, b) = 2$, then the distance from the midpoint, m, to a is 1 and $d(m, b) = 1$. Thus, $a = 4$ and $b = 6$.

54. Using the properties of inequalities, we have

$$\frac{1}{4}x - 3 < \frac{1}{2}x + 1$$
$$-\frac{1}{4}x + \left(\frac{1}{4}x - 3\right) - 1 < -\frac{1}{4}x + \left(\frac{1}{2}x + 1\right) - 1$$
$$-4 < \frac{1}{4}x$$
$$-16 < x$$
$$x > -16.$$

In interval notation this is $(-16, \infty)$.

57. The inequality $|x| > 10$ is equivalent to

$$x > 10 \qquad \text{or} \qquad x < -10.$$

In interval notation this is $(-\infty, -10) \cup (10, \infty)$.

60. The inequality $|5 - 2x| \geq 7$ is equivalent to

$$5 - 2x \geq 7 \qquad \text{or} \qquad 5 - 2x \leq -7.$$

Solving $5 - 2x \geq 7$ we obtain $-2x \geq 2$, so $x \leq -1$. In interval notation this is $(-\infty, -1]$. Solving $5 - 2x \leq -7$ we obtain $-2x \leq -12$, so $x \geq 6$. In interval notation this is $[6, \infty)$. The solution set is $(-\infty, 1] \cup [6, \infty)$.

63. Rewriting the inequality and factoring, we have

$$x^3 > x$$
$$x^3 - x > 0$$
$$x(x^2 - 1) > 0$$
$$x(x - 1)(x + 1) > 0.$$

Placing $x = -1$, $x = 0$, and $x = 1$ on the number line determines four intervals. The sign chart below, in turn, determines the solution set.

$x + 1$	$-$	0	$+$	$+$	$+$	$+$	$+$
$x - 1$	$-$	$-$	$-$	$-$	$-$	0	$+$
x	$-$	$-$	$-$	0	$+$	$+$	$+$
$x(x - 1)(x + 1)$	$-$	0	$+$	0	$-$	0	$+$

$$\xrightarrow[\quad -1 \qquad\qquad 0 \qquad\qquad 1 \quad]{\quad \big(\qquad\qquad \big)\qquad\qquad \big(\quad}$$

The solution set is $(-1, 0) \cup (1, \infty)$.

66. First, rewrite the inequality as a simple fraction:

$$\frac{2x - 6}{x - 1} > 1$$
$$\frac{2x - 6}{x - 1} - 1 > 0$$
$$\frac{2x - 6 - (x - 1)}{x - 1} > 0$$
$$\frac{x - 5}{x - 1} > 0.$$

Placing $x = 1$ and $x = 5$ on the number line determines three intervals. The sign chart below, in turn, determines the solution set.

$x - 1$	$-$	0	$+$	$+$	$+$
$x - 5$	$-$	$-$	$-$	0	$+$
$(x - 5)/(x - 1)$	$+$	undefined	$-$	0	$+$

$$\xleftarrow[\qquad 1 \qquad\qquad\qquad\qquad 5 \qquad]{\qquad\quad \big)\qquad\qquad\qquad\qquad \big(\qquad}\rightarrow$$

The solution set is $(-\infty, 1) \cup (5, \infty)$.

69. (a) Rationalizing the denominator, we have

$$\frac{x^2-16}{\sqrt{x}-2} = \frac{x^2-16}{\sqrt{x}-2}\frac{\sqrt{x}+2}{\sqrt{x}+2} = \frac{(x-4)(x+4)(\sqrt{x}+2)}{x-4} = (x+4)(\sqrt{x}+2).$$

(b) $\displaystyle\lim_{x\to4}\frac{x^2-16}{\sqrt{x}-2} = \lim_{x\to4}(x+4)(\sqrt{x}+2) = (4+4)(\sqrt{4}+2) = 8(4) = 32.$

Chapter 2

Functions

2.1 | Functions and Graphs |

3. $f(-1) = \sqrt{-1+1} = \sqrt{0} = 0$

$f(0) = \sqrt{0+1} = \sqrt{1} = 1$

$f(3) = \sqrt{3+1} = \sqrt{4} = 2$

$f(5) = \sqrt{5+1} = \sqrt{6}$

6. $f(-\sqrt{2}) = \dfrac{(-\sqrt{2})^2}{(-\sqrt{2})^3 - 2} = \dfrac{2}{-2\sqrt{2} - 2} = \dfrac{1}{-\sqrt{2} - 1} = -\dfrac{1}{\sqrt{2} + 1}$

$f(-1) = \dfrac{(-1)^2}{(-1)^3 - 2} = \dfrac{1}{-1 - 2} = \dfrac{1}{-3} = -\dfrac{1}{3}$

$f(0) = \dfrac{0^2}{0^3 - 2} = \dfrac{0}{-2} = 0$

$f\left(\dfrac{1}{2}\right) = \dfrac{\left(\dfrac{1}{2}\right)^2}{\left(\dfrac{1}{2}\right)^3 - 2} = \dfrac{\dfrac{1}{4}}{\dfrac{1}{8} - 2} = \dfrac{\dfrac{1}{4}}{\dfrac{1}{8} - 2}\left(\dfrac{8}{8}\right) = \dfrac{2}{1 - 16} = \dfrac{2}{-15} = -\dfrac{2}{15}$

9. Setting $f(x) = 23$ and solving for x, we find

$$6x^2 - 1 = 23$$
$$6x^2 = 24$$
$$x^2 = 4$$
$$x = \pm 2.$$

When we compute $f(-2)$ and $f(2)$ we obtain 23 in both cases, so $x = \pm 2$ is the answer.

12. The domain of $f(x) = \sqrt{15 - 5x}$ is the set of all x for which $15 - 5x \geq 0$. This is equivalent to $15 \geq 5x$, $3 \geq x$, and $x \leq 3$. The domain of $f(x)$ is $(-\infty, 3]$.

15. The domain of $f(x) = (2x - 5)/x(x - 3)$ is the set of all x for which $x(x - 3) \neq 0$. Since $x(x - 3) = 0$ when $x = 0$ or $x = 3$, the domain of $f(x)$ is $\{x | x \neq 0,\ x \neq 3\}$.

18. The domain of $f(x) = (x+1)/(x^2-4x-12)$ is the set of all x for which $x^2-4x-12 \neq 0$. Since $x^2 - 4x - 12 = (x+4)(x-6) = 0$ when $x = -4$ or $x = 6$, the domain of $f(x)$ is $\{x | x \neq -4, \ x \neq 6\}$.

21. The domain of $f(x) = \sqrt{25 - x^2}$ is the set of all x for which $25 - x^2 \geq 0$. We solve this inequality by writing $25 - x^2 = (5+x)(5-x) = 0$ and placing $x = -5$ and $x = 5$ on the number line. This determines three intervals, and the sign chart below then determines the solution set.

$5 + x$	$-$	0	$+$	$+$	$+$
$5 - x$	$+$	$+$	$+$	0	$-$
$(5+x)(5-x)$	$-$	0	$+$	0	$-$

The domain of the function is $[-5, 5]$.

24. The domain of $f(x) = \sqrt{x^2 - 3x - 10}$ is the set of all x for which $x^2 - 3x - 10 \geq 0$. We solve this inequality by writing $x^2 - 3x - 10 = (x+2)(x-5) = 0$ and placing $x = -2$ and $x = 5$ on the number line. This determines three intervals, and the sign chart below then determines the solution set.

$x + 2$	$-$	0	$+$	$+$	$+$
$x - 5$	$-$	$-$	$-$	0	$+$
$(x+2)(x-5)$	$+$	0	$-$	0	$+$

The domain of the function is $(-\infty, -2] \cup [5, \infty)$.

27. Since the y-axis (a vertical line) intersects the graph in more than one point (three points in this case), the graph is not that of a function.

30. Since the y-axis (a vertical line) intersects the graph in more than one point (three points in this case), the graph is not that of a function.

33. Horizontally, the graph extends between $x = 1$ and $x = 9$ and terminates at both ends, as indicated by the solid dots. Thus, the domain is $[1, 9]$. Vertically, the graph extends between $y = 1$ and $y = 6$, so the range is $[1, 6]$.

36. We solve $f(x) = -2x - 9 = 0$:

$$-2x + 9 = 0$$
$$-2x = -9$$
$$x = \frac{-9}{-2} = \frac{9}{2}.$$

Thus, the zero of the function is $x = \frac{9}{2}$.

39. Setting $f(x) = x(3x-1)(x+9) = 0$ we see that the zeros of the function occur where $x = 0$, $3x - 1 = 0$ or $x = \frac{1}{3}$, and $x = -9$.

42. We solve $f(x) = 2 - \sqrt{4 - x^2} = 0$:

$$2 - \sqrt{4 - x^2} = 0$$
$$2 = \sqrt{4 - x^2}$$
$$4 = 4 - x^2 \qquad \leftarrow \quad \text{square both sides of the equality}$$
$$-4 + 4 = -4 + 4 - x^2$$
$$0 = -x^2$$
$$x^2 = 0$$
$$x = 0.$$

This result is easily checked, verifying that $x = 0$ is a solution, so the only zero of f is 0.

45. *x-intercepts*: We solve $f(x) = 4(x-2)^2 - 1 = 0$:

$$4(x-2)^2 - 1 = 0$$
$$4(x-2)^2 = 1$$
$$(x-2)^2 = \frac{1}{4}$$
$$x - 2 = \pm\sqrt{\frac{1}{4}} = \pm\frac{1}{2}$$
$$x = 2 \pm \frac{1}{2}.$$

Both of these check with the original equation, so the x-intercepts are $\left(\frac{3}{2}, 0\right)$ and $\left(\frac{5}{2}, 0\right)$.

y-intercept: Since $f(0) = 4(0-2)^2 - 1 = 4(4) - 1 = 15$, the y-intercept is $(0, 15)$.

48. *x-intercept*: We solve $f(x) = x(x+1)(x-6)/(x+8) = 0$:

$$\frac{x(x+1)(x-6)}{x+8} = 0$$
$$x(x+1)(x-6) = 0, \quad x \neq -8.$$

Thus, $x = 0$, $x = -1$, and $x = 6$. The x-intercepts are $(0, 0)$, $(-1, 0)$, and $(6, 0)$.

y-intercept: Since $f(0) = 0$, the y-intercept is $(0, 0)$.

Note: Whenever the origin is an x-intercept of a function, it is also the y-intercept.

51. Solving $x = y^2 - 5$ for y, we obtain

$$x = y^2 - 5$$
$$x + 5 = y^2$$
$$y^2 = x + 5$$
$$y = \pm\sqrt{x+5}.$$

The two functions are $f_1(x) = -\sqrt{x+5}$ and $f_2(x) = \sqrt{x+5}$. The domains are both $[-5, \infty)$.

54. The function values are the directed distances from the x-axis at the given value of x.

$$f(-3) \approx 0; \quad f(-2) \approx -3.5; \quad f(-1) \approx 0.3; \quad f(1) \approx 2; \quad f(2) \approx 3.8; \quad f(3) \approx 2.8.$$

57. (a) $f(2) = 2! = 2 \cdot 1 = 2$

$f(3) = 3! = 3 \cdot 2 \cdot 1 = 6$

$f(5) = 5! = 5 \cdot 4 \cdot 3 \cdot 2 \cdot 1 = 120$

$f(7) = 7! = 7 \cdot 6 \cdot 5 \cdot 4 \cdot 3 \cdot 2 \cdot 1 = 5040$

Note that we could have simplified the computation of 7! in this case by writing

$$7! = 7 \cdot 6 \cdot 5! = 7 \cdot 6 \cdot 120 = 5040.$$

(b) $f(n+1) = (n+1)! = (n+1)n! = n!(n+1) = f(n)(n+1)$

(c) $\dfrac{f(n+2)}{f(n)} = \dfrac{(n+2)!}{n!} = \dfrac{(n+2)(n+1)n!}{n!} = (n+2)(n+1)$

2.2   Symmetry and Transformations

3. Since

$$f(-x) = (-x)^3 - (-x) + 4 = -x^3 + x + 4$$

is neither $f(x)$ nor $-f(x)$, the function is neither even nor odd.

6. Since

$$f(-x) = \frac{-x}{(-x)^2 + 1} = -\frac{x}{x^2 + 1} = -f(x),$$

the function is odd. Because the only function that is both even and odd is $f(x) = 0$, and f is odd in this case, it cannot also be even. [See **Even and Odd Functions** in the **Algebra Topics** part of this manual.]

9. Since

$$f(-x) = |(-x)^3| = |-x^3| = x^3 = f(x),$$

the function is even. Because the only function that is both even and odd is $f(x) = 0$, and f is even in this case, it cannot also be odd. [See **Even and Odd Functions** in the **Algebra Topics** part of this manual.]

12. The function is odd because we can see from the graph that $f(-x) = -f(x)$.

15. **(a)** An even function is symmetric with respect to the y-axis.

(b) An odd function is symmetric with respect to the origin.

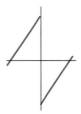

18. **(a)** An even function is symmetric with respect to the y-axis.

(b) An odd function is symmetric with respect to the origin.

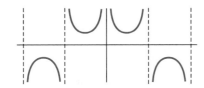

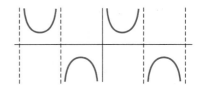

21. Since g is an odd function, $g(-x) = -g(x)$. Thus, taking $x = -1$, we have $g\big(-(-1)\big) = -g(-1) = -(-5) = 5$, and taking $x = 4$, we have $g(-4) = -g(4) = -8$.

24. Shifting $(2, 1)$ down 5 units we get $(2, 1 - 5)$ or $(2, -4)$. Shifting $(3, -4)$ down 5 units we get $(3, -4 - 5)$ or $(3, -9)$.

27. Shifting $(-2, 1)$ up 1 unit and left 4 units we get $(-2 - 4, 1 + 1)$ or $(-6, 2)$. Shifting $(3, -4)$ up 1 unit and left 4 units we get $(3 - 4, -4 + 1)$ or $(-1, -3)$.

30. Reflecting $(-2, 1)$ in the x-axis changes the y-coordinate from 1 to -1. The reflected point is then $(-2, -1)$. Reflecting $(3, -4)$ in the x-axis changes the y-coordinate from -4 to 4. The reflected point is then $(3, 4)$.

33. In **(a)** the graph is shifted up 2 units; in **(b)** it is shifted down 2 units; in **(c)** it is shifted left 2 units; in **(d)** it is shifted right 3 units; in **(e)** it is reflected in the x-axis; and in **(f)** it is reflected in the y-axis.

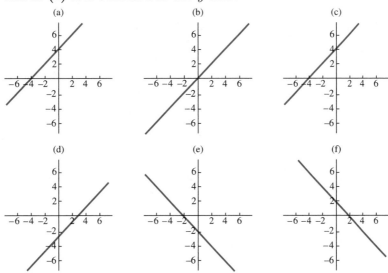

36. In **(a)** the graph is shifted up 2 units; in **(b)** it is shifted down 2 units; in **(c)** it is shifted left 2 units; in **(d)** it is shifted right 3 units; in **(e)** it is reflected in the x-axis; and in **(f)** it is reflected in the y-axis.

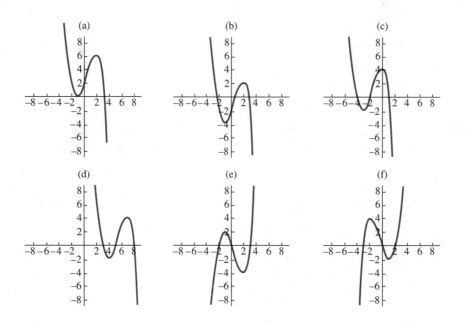

39. If $f(x)$ is shifted up 5 units and right 1 unit, the new function is $y = f(x-1) + 5$. Since $f(x) = x^3$, this becomes $y = (x-1)^3 + 5$.

42. If $f(x)$ is reflected in the y-axis, then shifted left 5 units and down 10 units, the new function is $y = -f(x+5) - 10$. Since $f(x) = 1/x$, this becomes $y = -1/(x+5) - 10$.

2.3   Linear Functions

3. The slope of the line through $(5, 2)$ and $(4, -3)$ is

$$m = \frac{-3 - 2}{4 - 5} = \frac{-5}{-1} = 5.$$

The graph is shown to the right.

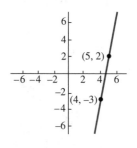

6. The slope of the line through $(8, -\frac{1}{2})$ and $(2, \frac{5}{2})$ is

$$m = \frac{\frac{5}{2} - \left(-\frac{1}{2}\right)}{2 - 8} = \frac{3}{-6} = -\frac{1}{2}.$$

The graph is shown to the right.

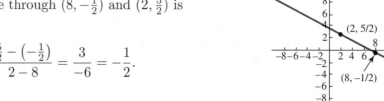

9. To find the x-intercept we set $y = 0$. This gives $3x + 12 = 0$ or $x = -4$. The x-intercept is $(-4, 0)$. Now, write the equation in slope-intercept form by solving for y:

$$3x - 4y + 12 = 0$$
$$-4y = -3x - 12$$
$$y = \frac{3}{4}x + 3.$$

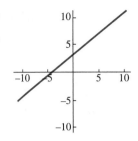

The slope of the line is $m = \frac{3}{4}$ and the y-intercept is $(0, 3)$. The graph of the line is shown to the right.

12. To find the x-intercept we set $y = 0$. This gives $-4x + 6 = 0$ or $x = \frac{3}{2}$. The x-intercept is $(\frac{3}{2}, 0)$. Now, write the equation in slope-intercept form by solving for y:

$$-4x - 2y + 6 = 0$$
$$-2y = 4x - 6$$
$$y = -2x + 3.$$

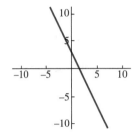

The slope of the line is $m = -2$ and the y-intercept is $(0, 3)$. The graph of the line is shown to the right.

15. To find the x-intercept we set $y = 0$. This gives $\frac{2}{3}x = 1$ or $x = \frac{3}{2}$. The x-intercept is $(\frac{3}{2}, 0)$. Now, write the equation in slope-intercept form by solving for y:

$$y + \frac{2}{3}x = 1$$
$$y = -\frac{2}{3}x + 1.$$

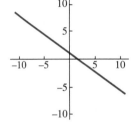

The slope of the line is $m = -\frac{2}{3}$ and the y-intercept is $(0, 1)$. The graph of the line is shown to the right.

18. Letting $m = \frac{1}{10}$, $x_1 = 1$, and $y_1 = 2$, we obtain from the point-slope form of the equation of a line that

$$y - 2 = \frac{1}{10}(x - 1)$$
$$y - 2 = \frac{1}{10}x - \frac{1}{10}$$
$$y = \frac{1}{10}x + \frac{19}{10}.$$

21. Letting $m = -1$, $x_1 = 1$, and $y_1 = 2$, we obtain from the point-slope form of the equation of a line that

$$y - 2 = -1(x - 1)$$
$$y - 2 = -x + 1$$
$$y = -x + 3.$$

24. The slope of the line is

$$m = \frac{-5 - 3}{6 - 2} = \frac{-8}{4} = -2,$$

and we identify $x_1 = 2$ and $y_1 = 3$. Then, using the point-slope form of the equation of a line, we have

$$y - 3 = -2(x - 2)$$
$$y - 3 = -2x + 4$$
$$y = -2x + 7.$$

27. Since the x-coordinates of the two points are the same, the points determine the vertical line $x = -2$.

30. Solving $2x - 5y + 4 = 0$ for y, we obtain

$$-5y = -2x - 4 \qquad \text{or} \qquad y = \frac{2}{5}x + \frac{4}{5}.$$

Thus, the slope of the parallel line is $\frac{2}{5}$. Now, using the point-slope form of the equation of a line, we identify $m = \frac{2}{5}$, $x_1 = 1$, and $y_1 = -3$, so that

$$y - (-3) = \frac{2}{5}(x - 1)$$
$$y + 3 = \frac{2}{5}x - \frac{2}{5}$$
$$y = \frac{2}{5}x - \frac{17}{5}.$$

33. Solving $x - 4y + 1 = 0$ for y, we obtain

$$-4y = -x - 1 \qquad \text{and} \qquad y = \frac{1}{4}x + \frac{1}{4}.$$

Thus, the slope of a line perpendicular to this one is $m = -1/(1/4) = -4$. We identify $x_1 = 2$ and $y_1 = 3$ and use the point-slope form of the equation of a line:

$$y - 3 = -4(x - 2)$$
$$y - 3 = -4x + 8$$
$$y = -4x + 11.$$

36. A line perpendicular to another line with slope 2 has slope $m = -\frac{1}{2}$. Since the line passes through the origin, the y-intercept is 0. Using the slope-intercept form of the equation of a line, we then have $y = -\frac{1}{2}x + 0$ or $y = -\frac{1}{2}x$.

39. We find the slopes of the given lines by solving the equation for y:

(a) $3x - y - 1 = 0$; $\quad -y = -3x + 1$; $\quad y = 3x - 1$; $\qquad m = 3$

(b) $x - 3y + 9 = 0$; $\quad -3y = -x - 9$; $\quad y = \frac{1}{3}x + 3$; $\qquad m = \frac{1}{3}$

(c) $3x + y = 0$; $\quad y = -3x$; $\qquad m = -3$

(d) $x + 3y = 1$; $3y = -x + 1$; $y = -\frac{1}{3}x + \frac{1}{3}$; $m = -\frac{1}{3}$

(e) $6x - 3y + 10 = 0$; $-3y = -6x - 10$; $y = 2x + \frac{10}{3}$; $m = 2$

(f) $x + 2y = -8$; $2y = -x - 8$; $y = -\frac{1}{2}x - 4$; $m = -\frac{1}{2}$

Parallel lines have equal slopes, so none of the lines is parallel to any of the other lines. The slopes of perpendicular lines are negative reciprocals of each other, so the lines in **(a)** and **(d)**, **(b)** and **(c)**, and **(e)** and **(f)** are perpendicular.

42. We are looking for a function having the form $f(x) = ax + b$. The first condition implies

$$a(-1) + b = 1 + a(2) + b$$
$$-a + b = 2a + b + 1$$
$$-3a = 1$$
$$a = -\frac{1}{3}.$$

The second condition implies

$$a(3) + b = 4[a(1) + b]$$
$$3a + b = 4a + 4b$$
$$0 = a + 3b.$$

Since we have already determined $a = -\frac{1}{3}$,

$$0 = -\frac{1}{3} + 3b$$
$$\frac{1}{3} = 3b$$
$$b = \frac{1}{9}.$$

The linear function is $f(x) = -\frac{1}{3}x + \frac{1}{9}$.

45. The lines are $y = 4x + 7$ and $y = \frac{1}{3}x + \frac{10}{3}$. We equate these and solve for x:

$$4x + 7 = \frac{1}{3}x + \frac{10}{3}$$
$$\left(4 - \frac{1}{3}\right)x = \frac{10}{3} - 7$$
$$\frac{11}{3}x = -\frac{11}{3}$$
$$x = -1.$$

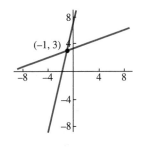

Substituting $x = -1$ into $y = 4x + 7$ we find $y = 4(-1) + 7 = -4 + 7 = 3$. The point of intersection of the lines is $(-1, 3)$. The graph is shown to the right. *Note*: This answer is easily checked. Simply plug $x = -1$ and $y = 3$ into the other equation, $y = \frac{1}{3}x + \frac{10}{3}$ to confirm that the point also lies on this line. Since $\frac{1}{3}(-1) + \frac{10}{3} = -\frac{1}{3} + \frac{10}{3} = \frac{9}{3} = 3$, you can be fairly confident that the point of intersection is actually $(-1, 3)$.

48. We first compute

$$f(x+h) = \frac{4}{3}(x+h) - 5 = \frac{4}{3}x + \frac{4}{3}h - 5.$$

Then

$$\frac{f(x+h) - f(x)}{h} = \frac{\frac{4}{3}x + \frac{4}{3}h - 5 - \left(\frac{4}{3}x - 5\right)}{h} = \frac{\frac{4}{3}h}{h} = \frac{4}{3}.$$

51. (a) Since T_F is a linear function of T_C, we have $T_F = aT_C + b$, for appropriate choices of a and b. Since $T_F = 32$ when $T_C = 0$ and $T_F = 140$ when $T_C = 60$, we have $32 = a(0) + b = b$ and $140 = a(60) + b$. Then

$$140 = 60a + 32; \quad 60a = 108; \quad \text{and} \quad a = \frac{108}{60} = \frac{9}{5}.$$

Thus, $T_F = \frac{9}{5}T_C + 32$.

(b) When $T_C = 100$ we have $T_F + \frac{9}{5}(100) + 32 = 180 + 32 = 212$.

54. If x is the number of years, then $A(x) = ax + b$ for appropriate choices of a and b. Since $A = 20{,}000$ when $x = 0$, we have $20{,}000 = a(0) + b = b$. Since $A = 0$ when $x = 25$, we have $0 = a(25) + b = 25a + 20{,}000$, so $25a = -20{,}000$ and $a = -400$. Thus, $A(x) = -400x + 20{,}000$.

2.4   Quadratic Functions

3. The graph of $f(x) = 2x^2 - 2$ is a parabola that opens upward because the coefficient of x^2 is positive. Identifying $a = 2$ and $b = 0$, we see that the vertex is at $x = -b/2a = 0$. Thus, the vertex is $\big(0, f(0)\big) = (0, -2)$. To find the x-intercepts we solve $2x^2 - 2 = 0$:

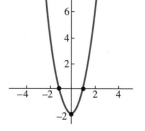

$$x^2 - 1 = 0$$
$$x^2 = 1$$
$$x = \pm 1.$$

Thus, $(-1, 0)$ and $(1, 0)$ are on the graph. Plotting the three points, $(-1, 0)$, $(0, -2)$, and $(1, 0)$, we obtain the graph of the parabola.

6. The graph of $f(x) = -2x^2 - 3$ is a parabola that opens downward because the coefficient of x^2 is negative. Identifying $a = -2$ and $b = 0$ we see that the vertex is at $x = -b/2a = 0$. Thus, the vertex is $\big(0, f(0)\big) = (0, -3)$. Since $-2x^2 - 3 = 0$ is equivalent to $x^2 = -\frac{3}{2}$, the equation has no solutions and there are no x-intercepts. In this case, the only intercept is the vertex. To graph the function we need another point on the graph. To obtain such a point simply choose any x other than $x = 0$ and find $f(x)$. For example, if we let $x = 2$, we obtain $f(2) = -2(2)^2 - 3 = -2(4) - 3 = -11$. Thus, the point $(2, -11)$ is on the graph. Since f is an even function, $(-2, -11)$ is also on the graph and we plot the parabola through $(-2, -11)$, $(0, -3)$, and $(2, -11)$.

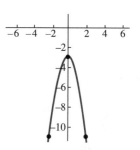

9. **(a)** Solving $(3-x)(x+1) = 0$ we obtain $x = 3$ and $x = -1$, so the x-intercepts are $(-1,0)$ and $(3,0)$. The y-intercept is at $f(0) = (3-0)(0+1) = 3$, so it is $(0,3)$.

   (b) We expand the expression and complete the square:

   $$
   \begin{aligned}
   f(x) &= (3-x)(x+1) \\
   &= -x^2 + 2x + 3 \\
   &= -(x^2 - 2x \quad) + 3 \\
   &= -(x^2 - 2x + 1) + 3 + 1 \\
   &= -(x-1)^2 + 4.
   \end{aligned}
   $$

   (c) Identifying $h = 1$ and $k = 4$ we see that the vertex is $(1,4)$. The axis of symmetry is $x = 1$.

   (d) Since the coefficient of x^2 is negative, the graph opens downward. We use the vertex and intercepts to draw the graph.

   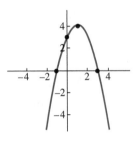

12. **(a)** Factoring, we obtain

   $$f(x) = -x^2 + 6x - 5 = -(x^2 - 6x + 5) = -(x-1)(x-5).$$

   Solving $-(x-1)(x-5) = 0$ we obtain $x = 1$ and $x = 5$, so the x-intercepts are $(1,0)$ and $(5,0)$. The y-intercept is at $f(0) = -5$, so it is $(0,-5)$.

   (b) To complete the square and obtain the standard form, we start by factoring -1 from the two x-terms:

   $$
   \begin{aligned}
   f(x) &= -x^2 + 6x - 5 \\
   &= -(x^2 - 6x \quad) - 5 \\
   &= -(x^2 - 6x + 9) - 5 + 9 \\
   &= -(x-3)^2 + 4.
   \end{aligned}
   $$

   (c) Identifying $h = 3$ and $k = 4$ we see that the vertex is $(3,4)$. The axis of symmetry is $x = 3$.

   (d) Since the coefficient of x^2 is negative, the graph opens downward. We use the vertex and intercepts to draw the graph.

   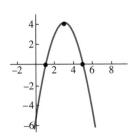

15. **(a)** Factoring, we obtain

$$f(x) = -\frac{1}{2}x^2 + x + 1 = -\frac{1}{2}(x^2 - 2x - 2).$$

Using the quadratic formula we find that the roots of $x^2 - 2x - 2$ are $(2 \pm \sqrt{4 + 8})/2 = 1 \pm \sqrt{3}$, so the x-intercepts are $(1 - \sqrt{3}, 0)$ and $(1 + \sqrt{3}, 0)$. The y-intercept is at $f(0) = 1$, so the y-intercept is $(0, 1)$.

(b) To complete the square and obtain the standard form we start by factoring $-\frac{1}{2}$ from the two x-terms:

$$
\begin{aligned}
f(x) &= -\frac{1}{2}x^2 + x + 1 \\
&= -\frac{1}{2}(x^2 - 2x \quad) + 1 \\
&= -\frac{1}{2}(x^2 - 2x + 1) + 1 + \frac{1}{2}(1) \\
&= -\frac{1}{2}(x - 1)^2 + \frac{3}{2}.
\end{aligned}
$$

(c) Identifying $h = 1$ and $k = \frac{3}{2}$ we see that the vertex is $(1, \frac{3}{2})$. The axis of symmetry is $x = 1$.

(d) Since the coefficient of x^2 is negative, the graph opens downward. We use the vertex and intercepts to draw the graph.

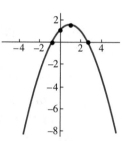

18. **(a)** Factoring, we obtain

$$f(x) = -x^2 + 6x - 9 = -(x^2 - 6x + 9) = -(x - 3)^2.$$

Setting $f(x) = -(x - 3)^2 = 0$ we obtain $x = 3$, so the x-intercept is $(3, 0)$. The y-intercept is at $f(0) = -9$, so it is $(0, -9)$.

(b) The function in standard form is $f(x) = -(x - 3)^2$.

(c) Identifying $h = 3$ and $k = 0$ we see that the vertex is $(3, 0)$. The axis of symmetry is $x = 3$.

(d) Since the coefficient of x^2 is negative, the graph opens downward. We use the vertex and intercept to draw the graph.

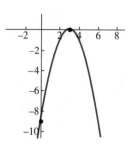

21. The function is in standard form, so we identify $h = 0$ and $k = -25$. Since the coefficient of x^2 is positive, the parabola opens upward and the function is decreasing on $(-\infty, 0]$ and increasing on $[0, \infty)$.

24. To put the function in standard form we complete the square:

$$\begin{aligned} f(x) &= x^2 + 8x - 1 \\ &= (x^2 + 8x \qquad) - 1 \\ &= (x^2 + 8x + 16) - 1 - 16 \\ &= (x + 4)^2 - 17. \end{aligned}$$

We identify $h = -4$. Since the coefficient of x^2 is positive, the parabola opens upward and the function is decreasing on $(-\infty, -4]$ and increasing on $[-4, \infty)$.

27. The vertex of the graph of $f(x) = -\frac{1}{3}(x + 4)^2 + 9$ is $(-4, 9)$, so the graph of $f(x)$ is the graph of $y = x^2$ reflected in the x-axis, compressed by a factor of $\frac{1}{3}$, shifted to the left by 4 units, and shifted up 9 units.

30. Since $(a - b)^2 = (b - a)^2$ for any choice of a and b, the vertex of the graph of

$$f(x) = -(1 - x)^2 + 1 = -(x - 1)^2 + 1$$

is $(1, -1)$. Thus, the graph of $f(x)$ is the graph of $y = x^2$ reflected in the x-axis, shifted to the right by 1 unit, and shifted up 1 unit.

33. The graph of $y = x^2$ has been reflected in the x-axis and shifted down by 1 unit. Thus, its equation is $y = -x^2 - 1$.

36. The graph of $y = x^2$ has been reflected in the x-axis and shifted up by 3 units. Thus, its equation is $y = -x^2 + 3$.

39. Since the vertex of f is $(1, 2)$, we identify $h = 1$ and $k = 2$. Then, the equation is $f(x) = a(x - 1) + 2$. Since

$$f(2) = a(2 - 1) + 2 = a + 2 = 6,$$

we have $a = 4$ and $f(x) = 4(x - 1)^2 + 2$.

42. To find the points of intersection of the two graphs, we solve $2x - 2 = 1 - x^2$, since both sides of this equation are equal to y:

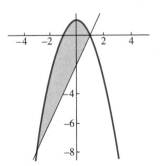

$$\begin{aligned} 2x - 2 &= 1 - x^2 \\ x^2 + 2x - 3 &= 0 \\ (x + 3)(x - 1) &= 0. \end{aligned}$$

When $x = -3$, $y = 2(-3) - 2 = -8$ and when $x = 1$, $y = 2(1) - 2 = 0$, so the points of intersection are $(-3, -8)$ and $(1, 0)$. The region between the graphs is shown in the figure.

45. (a) The point on the graph is $(x, y) = (x, 2x)$, so the distance from $(5, 0)$ to this point is

$$d = \sqrt{(x-5)^2 + (2x-0)^2} = \sqrt{x^2 - 10x + 25 + 4x^2} = \sqrt{5x^2 - 10x + 25}\,.$$

(b) The minimum value of d will occur where $5x^2 - 10x + 25$ is minimum. The graph of $y = 5x^2 - 10x + 25$ is a parabola that opens upward, so the minimum value of y will occur at the vertex of the parabola. To find the vertex we put the equation in standard form:

$$\begin{aligned}
y &= 5x^2 - 10x + 25 \\
&= 5(x^2 - 2x \qquad) + 25 \\
&= 5(x^2 - 2x + 1) + 25 - 5 \\
&= 5(x-1)^2 + 20.
\end{aligned}$$

We see that the vertex is at $(1, 20)$, so d is a minimum when $x = 1$. The corresponding point on the line $y = 2x$ is $2(1) = 2$. Thus, the point on the line $y = 2x$ closest to $(5, 0)$ is $(1, 2)$.

48. (a) As discussed in the text, the height $s(t)$ of an object at time t that is thrown or dropped from an initial height s_0 with an initial velocity v_0 is given by

$$s(t) = \frac{1}{2}gt^2 + v_0 t + s_0.$$

The value of g in this formula depends on the units that are used to measure distance. Since we are given $s(t) = -16t^2 + 96t + 256$, we see that $g = -32$, so the units of distance are feet, and the initial height is $s_0 = 256$ feet. The height of the building is then 256 feet.

(b) The trajectory of the rocket is a parabola in the s, t-plane and the maximum height attained by the rocket is the s-coordinate of the vertex. To find the vertex we express the quadratic equation for s in standard form by completing the square. We begin by factoring -16 from the two terms containing t:

$$\begin{aligned}
s &= -16t^2 + 96t + 256 \\
&= -16(t^2 - 6t \qquad) + 256 \\
&= -16(t^2 - 6t + 9) + 256 + 144 \\
&= -16(t-3)^2 + 400.
\end{aligned}$$

We see that the vertex of the parabola is $(3, 400)$, so the maximum height attained by the rocket is 400 feet.

(c) When the rocket strikes the ground, $s = 0$, so we solve

$$-16t^2 + 96t + 256 = 0$$

for t. Factoring, we have

$$-16t^2 + 96t + 256 = -16(t^2 - 6t - 16) = -16(t + 2)(t - 8),$$

so $t = -2$ and $t = 8$. We ignore $t = -2$ since the rocket is launched at time $t = 0$ and so t cannot be negative. We conclude that the rocket strikes the ground at $t = 8$ seconds. The graph of the portion of the parabola that is applicable to this problem is shown.

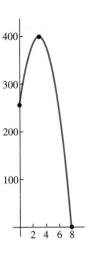

51. (a) To find when the disease spreads most rapidly, we want to determine when the rate of spread $R(D)$ is a maximum. That is, we want to find the value of D for which

$$R = kD(P - D) = kPD - kD^2$$

is a maximum. (Here, "when" does not refer to time, but rather to the number D of infected persons.) The equation $R = kPD - kD^2$, where k and P are constants, is quadratic, so its graph in the DR-plane is a parabola. The maximum value of R will be at the vertex of the parabola. To determine the vertex we put the equation in standard form by completing the square:

$$\begin{aligned}
R &= kPD - kD^2 \\
&= -k(D^2 - PD \qquad) \\
&= -k\left(D^2 - PD + \frac{1}{4}P^2\right) + \frac{1}{4}kP^2 \\
&= k\left(D - \frac{1}{2}P\right)^2 + \frac{1}{4}kP^2.
\end{aligned}$$

We see that the vertex is $(\frac{1}{2}P, \frac{1}{4}kP^2)$. Since the parabola opens downward (because $k > 0$), the maximum value of R occurs when $D = \frac{1}{2}P$. That is, when the number D of people carrying the disease is one-half the population P.

(b) Identifying $P = 10{,}000$, $D = 125$, and $R = 37$ we solve

$$37 = k(125)(10{,}000 - 125) = k(1{,}234{,}375)$$

for k. This gives $k = 37/1{,}234{,}375 \approx 0.00003$.

(c) The number of new cases on Tuesday will be

$$\begin{aligned}
R = 0.00003D(P - D) &= 0.00003(125 + 37)[10{,}0000 - (125 + 37)] \\
&= 0.00003(162)(9{,}838) \approx 48.
\end{aligned}$$

(d) Repeated use of the formula $R = kD(P - D)$ gives the following table showing the number of new cases each day.

Day	New Cases – R	Total Infected – D
Sunday		125
Monday	37	162
Tuesday	48	210
Wednesday	62	272
Thursday	79	351
Friday	102	453
Saturday	130	

(See Part II of this *Student Resource Manual* for a discussion on how a TI calculator can be used to generate the data in the table.)

2.5 Piecewise-Defined Functions

3. Since $1 \geq 1$, $f(1) = 1^2 + 2(1) = 1 + 2 = 3$.

Since $0 < 1$, $f(0) = -0^3 = 0$.

Since $-2 < 1$, $f(-2) = -(-2)^3 = -(-8) = 8$.

Since $\sqrt{2} \geq 1$, $f(\sqrt{2}) = (\sqrt{2})^2 + 2\sqrt{2} = 2 + 2\sqrt{2}$.

6. The y-intercept occurs where $x = 0$. In this case, since 0 is a rational number, $f(0) = 1$, and the y-intercept is $(0, 1)$.

9. The y-intercept is at $f(0) = -0 = 0$, so the y-intercept is $(0, 0)$.

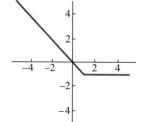

To find the x-intercepts, we set each part of the function equal to 0, solve for x, and then check to see if the value of x is in the interval determined by the appropriate part of the function:

$$-x = 0 \quad \text{so} \quad x = 0.$$

Since $0 \leq 1$, $(0, 0)$ is an x-intercept. (We could also have determined that it is an x-intercept because whenever the origin is the y-intercept, it must also be an x-intercept.)

Because -1 is never 0, there is no x-intercept for $x > 0$.

Since the graph has no holes, the function is continuous.

12. The y-intercept is at $f(0) = 0$, so the y-intercept is $(0, 0)$.

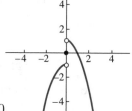

To find the x-intercepts, we set each part of the function equal to 0, solve for x, and then check to see if the value of x is in the interval determined by the appropriate part of the function:

$$-x^2 - 1 = 0 \quad \text{so} \quad x^2 = -1.$$

Since x^2 must be nonnegative, there is no x-intercept for $x < 0$.

Since $(0, 0)$ is the y-intercept, it must also be an x-intercept.

Setting $x^2 + 1 = 0$, we have $x^2 = -1$, so there is no x-intercept for $x > 0$.

The function is discontinuous at every integer value of x.

15. The y-intercept is at $f(0) = 0$, so the y-intercept is $(0,0)$.

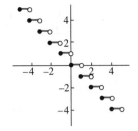

To find the x-intercepts, we solve $-[\![x]\!] = 0$. This is equivalent to $[\![x]\!] = 0$, and we see from Figure 2.5.3 in the text that $[\![x]\!] = 0$ for all x in the interval $[0,1)$. Thus, the x-intercepts are $[a,0)$ for $0 \le a < 1$.

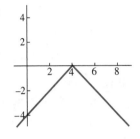

The function is discontinuous at every integer value of x.

18. The y-intercept is at $f(0) = -|0 - 4| = -4$, so the y-intercept is $(0,-4)$.

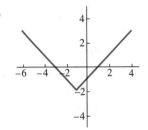

To find the x-intercepts, we solve $-|x - 4| = 0$. This is equivalent to $x = 4$, so the x-intercept is $(4,0)$.

Since the graph has no holes, the function is continuous.

21. The y-intercept is at $f(0) = -2 + |0 + 1| = -2 + 1 = -1$, so the y-intercept is $(0,-1)$.

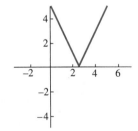

To find the x-intercepts we solve $-2 + |x + 1| = 0$:

$$-2 + |x + 1| = 0$$
$$|x + 1| = 2$$
$$x + 1 = \pm 2$$
$$x = -1 \pm 2.$$

Thus, $x = -1 - 2 = -3$ and $x = -1 + 2 = 1$. The x-intercepts are $(-3,0)$ and $(1,0)$.

Since the graph has no holes, the function is continuous.

24. The y-intercept is at $f(0) = |2(0) - 5| = 5$, so the y-intercept is $(0,5)$.

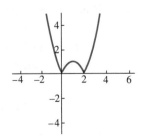

To find the x-intercepts we solve

$$|2x - 5| = 0 \quad \text{or} \quad 2x - 5 = 0.$$

Thus, $x = \frac{5}{2}$ and the x-intercept is $\left(\frac{5}{2},0\right)$.

Since the graph has no holes, the function is continuous.

27. The y-intercept is at $f(0) = |0^2 - 2(0)| = 0$, so the y-intercept is $(0,0)$.

Since the origin is a y-intercept, it is also an x-intercept. To find any other x-intercepts we set $|x^2 - 2x| = 0$. This is equivalent to $x^2 - 2x = x(x - 2) = 0$. Thus, a second x-intercept occurs at $x = 2$ and the x-intercepts are $(0,0)$ and $(2,0)$.

Since the graph has no holes, the function is continuous.

30. The y-intercept is at $f(0) = |\sqrt{0} - 2| = 2$, so the y-intercept is $(0, 2)$.

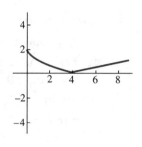

To find the x-intercepts we solve $|\sqrt{x} - 2| = 0$:

$$|\sqrt{x} - 2| = 0$$
$$\sqrt{x} - 2 = 0$$
$$\sqrt{x} = 2$$
$$x = 4.$$

Thus, the x-intercept is $(4, 0)$.

Since the graph has no holes, the function is continuous.

33. Since 0 is in the interval determined by $0 \leq x \leq 2$, the y-intercept is at $f(0) = |0 - 1| = 1$, so the y-intercept is $(0, 1)$.

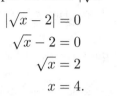

Since $y = 1$ for $x < 0$ and for $x > 2$, there can be no x-intercepts in these intervals. Setting $|x - 1| = 0$, we have $x - 1 = 0$, so the x-intercept is $(1, 0)$.

Since the graph has no holes, the function is continuous.

36. The graphs are shown below with the graph of $y = 2\,[\![x]\!]$ on the left and the graph of $y = [\![2x]\!]$ on the right. The scales on the two graphs are the same so it is easy to compare them. The steps on the graph of $y = 2\,[\![x]\!]$ are longer and further apart than those on the graph of $y = [\![2x]\!]$.

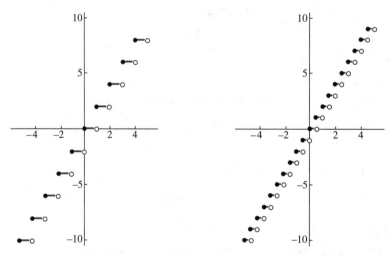

39. The line in the second quadrant is a portion of the graph of $y = -x$, while the graph of the line in the first quadrant is the graph of $y = x$. The semicircle is the graph of the upper half of the circle whose equation is $x^2 + y^2 = 9$, so this portion of the function is defined by $y = \sqrt{9 - x^2}$. The piecewise-defined function is

$$f(x) = \begin{cases} -x, & x < -3 \\ \sqrt{9 - x^2}, & -3 \leq x < 3 \\ x, & x \geq 3, \end{cases}$$

where the use of $<$ versus $\leq$ is based on the appearance of a circle versus a solid dot in the graph.

42. To graph the absolute value of the function determined by the given graph, simply reflect through the x-axis any portions of the graph that are below the x-axis.

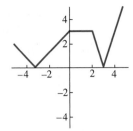

$$f(x) = \begin{cases} -x - 3, & x \le -3 \\ x + 3, & -3 < x \le 0 \\ 3, & 0 < x \le 2 \\ -3x + 9, & 2 < x \le 3 \\ 3x - 9, & 3 < x \end{cases}$$

45. Since $2 \le 2$, $f(2) = \frac{1}{2}(2) + 1 = 1 + 1 = 2$. We need to find k so that $k(2) = f(2) = 2$. This is equivalent to $2k = 2$ or $k = 1$.

48.

2.6   Combining Functions

3. $(f + g)(x) = f(x) + g(x) = x + \sqrt{x - 1}$

$(f - g)(x) = f(x) - g(x) = x - \sqrt{x - 1}$

$(fg)(x) = f(x)g(x) = x\sqrt{x - 1}$

$\left(\dfrac{f}{g}\right)(x) = \dfrac{f(x)}{g(x)} = \dfrac{x}{\sqrt{x - 1}}$

The domain of f is $(-\infty, \infty)$ and the domain of g is $\{x \mid x - 1 \ge 0\} = \{x \mid x \ge 1\}$ or $[1, \infty)$. The intersection of these sets is $(-\infty, \infty) \cap [1, \infty)$. Thus, the domains of $f + g$, $f - g$ and fg are all $[1, \infty)$.

The domain of g also includes the provision that $\sqrt{x - 1} \ne 0$. This is equivalent to $x - 1 \ne 0$ or $x \ne 1$. Thus, the domain of f/g is $(1, \infty)$.

6. $(f + g)(x) = f(x) + g(x) = \dfrac{4}{x - 6} + \dfrac{x}{x - 3} = \dfrac{4(x - 3) + x(x - 6)}{(x - 6)(x - 3)} = \dfrac{x^2 - 2x - 12}{x^2 - 9x + 18}$

$(f - g)(x) = f(x) - g(x) = \dfrac{4}{x - 6} - \dfrac{x}{x - 3} = \dfrac{4(x - 3) - x(x - 6)}{(x - 6)(x - 3)} = -\dfrac{x^2 - 10x + 12}{x^2 - 9x + 18}$

$(fg)(x) = f(x)g(x) = \dfrac{4}{x - 6}\left(\dfrac{x}{x - 3}\right) = \dfrac{4x}{x^2 - 9x + 18}$

$\left(\dfrac{f}{g}\right)(x) = \dfrac{f(x)}{g(x)} = \dfrac{4}{x - 6} \bigg/ \dfrac{x}{x - 3} = \dfrac{4(x - 3)}{x(x - 6)} = \dfrac{4x - 12}{x^2 - 6x}$

The domain of f is $\{x \mid x \ne 6\}$ and the domain of g is $\{x \mid x \ne 3\}$. The intersection of these sets is $(-\infty, 3) \cup (3, 6) \cup (6, \infty)$. Thus, the domains of $f + g$, $f - g$ and fg are all $(-\infty, 3) \cup (3, 6) \cup (6, \infty)$.

The domain of g also includes the provision that $x/(x - 3) \neq 0$. This is equivalent to $x \neq 0$. Thus, the domain of f/g is $(-\infty, 0) \cup (0, 3) \cup (3, 6) \cup (6, \infty)$. [Note that formula for f/g,

$$\left(\frac{f}{g}\right)(x) = \frac{4x - 12}{x^2 - 6x},$$

does not require that $x \neq 3$. Nevertheless, the value $x = 3$ must still be excluded from the domain of f/g because $g(x)$ is not defined for $x = 3$.]

9. To fill in the bottom row we compute

$$(f \circ g)(0) = f(g(0)) = f(2) = 10$$
$$(f \circ g)(1) = f(g(1)) = f(3) = 8$$
$$(f \circ g)(2) = f(g(2)) = f(0) = -1$$
$$(f \circ g)(3) = f(g(3)) = f(1) = 2$$
$$(f \circ g)(4) = f(g(4)) = f(4) = 0$$

x	0	1	2	3	4
$f(x)$	-1	2	10	8	0
$g(x)$	2	3	0	1	4
$(f \circ g)(x)$	10	8	-1	2	0

12. $(f \circ g)(x) = f(g(x)) = f(-x+4) = (-x+4)^2 - (-x+4) + 5 = x^2 - 8x + 16 + x - 4 + 5$

$$= x^2 - 7x + 17$$

$(g \circ f)(x) = g(f(x)) = g(x^2 - x + 5) = -(x^2 - x + 5) + 4 = -x^2 + x - 1$

The domains of $f \circ g$ and $g \circ f$ are both $(-\infty, \infty)$.

15. $(f \circ g)(x) = f(g(x)) = f(\frac{1}{2}(x + 3)) = 2[\frac{1}{2}(x + 3)] - 3 = x + 3 - 3 = x$

$(g \circ f)(x) = g(f(x)) = g(2x - 3) = \frac{1}{2}[(2x - 3) + 3] = \frac{1}{2}(2x) = x$

18. $(f \circ g)(x) = f(g(x)) = f(x^2) = \sqrt{x^2 - 4}$

$(g \circ f)(x) = g(f(x)) = g(\sqrt{x - 4}) = (\sqrt{x - 4})^2 = x - 4, \quad x \geq 4$

(Since $f(x) = \sqrt{x - 4}$ is only defined for $x - 4 \geq 0$ or $x \geq 4$, the domain of $g \circ f$ must be restricted to $x \geq 4$, even though the final expression $x - 4$ is defined for all x.)

21. $(f \circ f)(x) = f(f(x)) = f(2x + 6) = 2(2x + 6) + 6 = 4x + 18$

$$\left(f \circ \frac{1}{f}\right)(x) = f\left(\frac{1}{f}(x)\right) = f\left(\frac{1}{f(x)}\right) = f\left(\frac{1}{2x + 6}\right) = 2\left(\frac{1}{2x + 6}\right) + 6$$

$$= \frac{1}{x + 3} + 6 = \frac{1 + 6(x + 3)}{x + 3} = \frac{6x + 19}{x + 3}$$

24.

$$(f \circ f)(x) = f\big(f(x)\big) = f\left(\frac{x+4}{x}\right) = \frac{\dfrac{x+4}{x}+4}{\dfrac{x+4}{x}} = \frac{\dfrac{x+4}{x}+4}{\dfrac{x+4}{x}}\left(\frac{x}{x}\right) = \frac{x+4+4}{x+4}$$

$$= \frac{x+8}{x+4}$$

$$\left(f \circ \frac{1}{f}\right)(x) = f\left(\frac{1}{f}(x)\right) = f\left(\frac{1}{f(x)}\right) = f\left(\frac{1}{\dfrac{x+4}{x}}\right) = f\left(\frac{x}{x+4}\right)$$

$$= \frac{\dfrac{x}{x+4}+4}{\dfrac{x}{x+4}} = \frac{\dfrac{x}{x+4}+4}{\dfrac{x}{x+4}}\left(\frac{x+4}{x+4}\right) = \frac{x+4(x+4)}{x} = \frac{5x+16}{x}$$

27. $(f \circ g \circ g)(x) = f\big(g(g(x))\big) = f\big(g(3x^2)\big) = f\big(3(3x^2)^2\big) = f\big(3(9x^4)\big)$

$$= f(27x^4) = 2(27x^4) + 7 = 54x^4 + 7$$

30. $(f \circ f \circ f)(x) = f\big(f(f(x))\big) = f\big(f(x^2-1)\big) = f\big((x^2-1)^2-1\big) = f(x^4-2x^2+1-1)$

$$= f(x^4 - 2x^2) = (x^4 - 2x^2)^2 - 1 = x^8 - 4x^6 + 4x^4 - 1$$

33. In $f \circ g$ the first function to operate on x is g. One way to determine a choice for g is to ask yourself "if I were evaluating the function $F(x) = (x-3)^2 + 4\sqrt{x-3}$ at a specific number, say 7, what would be the first thing I would do with the 7?" You could answer this question by saying that you would first subtract 3 from the 7. In this case, then, $g(x) = x - 3$. Then, what is the next thing you would do? This would be to square the $x - 3$ and add it to 4 times the square root of $x - 3$. Thus, $f(x) = x^2 + 4x$. (Note that x is used here, not $x - 32$.)

36. First, we find formulas for $f \circ g$ and $g \circ f$:

$$(f \circ g)(x) = f\big(g(x)\big) = f(|x|) = [\![|x| - 1]\!]$$
$$(g \circ f)(x) = g\big(f(x)\big) = g([\![x-1]\!]) = \big|[\![x-1]\!]\big|.$$

The graphs are shown here.

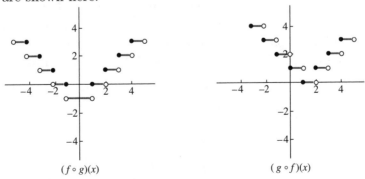

$(f \circ g)(x)$ $(g \circ f)(x)$

One way to obtain these graphs is to plot some points for values of x like

$$-3, \quad -2.5, \quad -2, \quad -1.5, \quad -1, \quad -0.5, \quad 0, \quad 0.5, \quad 1.5, \quad \text{and} \quad 2.5.$$

39. The graphs of $f(x) = |x - 1|$ and $g(x) = |x|$ are shown together with the graph of $(f + g)(x)$, which is shown with a blue curve. In this case, it is helpful to plot points at values of x like 0, 0.5, 1, 0.75, 2, and 3. Note also from the graphs of f and g, that the graph of $f + g$ is symmetric around the line $x = \frac{1}{2}$.

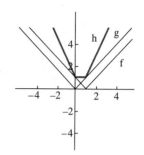

42. The graphs of $f(x) = x$ and $g(x) = [\![x]\!]$ are shown together with the graph of $(fg)(x)$, which is shown with a blue curve. Here you could discern the graph of fg by plotting points at values of x like 0, 0.5, 1, 1.5, 1.9, 2, 2.9, 3, -1, -1.1, and -2.

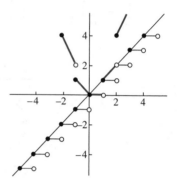

45. **(a)** Solving $x^2 - 1 = 1 - x$ we have

$$x^2 + x - 2 = 0$$
$$(x - 1)(x + 2) = 0$$

and thus,

$$x = 1, -2.$$

When $x = 1$, $y = 1 - 1 = 0$ (or $y = 1^2 - 1 = 0$), and when $x = -2$, $y = 1 - (-2) = 3$ (or $y = (-2)^2 - 1 = 3$). The points of intersection are $(1, 0)$ and $(-2, 3)$.

(b) To find d we subtract:

$$(1 - x) - (x^2 - 1) = -x^2 - x + 2.$$

Thus, $d(x) = -x^2 - x + 2$.

(c) The graph of $d(x)$ is a parabola opening downward (because the coefficient of x^2 is negative). Its vertex thus represents both the value at which the maximum value of $d(x)$ occurs as well as the actual maximum value of d. To find the vertex we express the quadratic function in standard form by completing the square:

$$\begin{aligned}
d &= -x^2 - x + 2 \\
&= -(x^2 + x \quad\quad) + 2 \\
&= -\left(x^2 + x + \frac{1}{4}\right) + 2 + \frac{1}{4} \\
&= -\left(x + \frac{1}{2}\right)^2 + \frac{9}{4}.
\end{aligned}$$

The vertex is $(-\frac{1}{2}, \frac{9}{4})$. We thus see that the maximum value of d occurs when $x = -\frac{1}{2}$ and the maximum value of d is $\frac{9}{4}$.

48. (a) The area of a circle is given by $A = \pi r^2$, so

$$A(t) = \pi[r(t)]^2 = \pi\left(4 - \frac{4}{t^2+1}\right)^2 = 4^2\pi\left(1 - \frac{1}{t^2+1}\right)^2 = 16\pi\left(\frac{t^2+1-1}{t^2+1}\right)^2$$

$$= 16\pi\left(\frac{t^2}{t^2+1}\right)^2 = 16\pi\,\frac{t^4}{(t^2+1)^2}.$$

(b) The circumference C of a circle is given by $C = 2\pi r$, so

$$C(t) = 2\pi r(t) = 2\pi\left(4 - \frac{4}{t^2+1}\right) = 8\pi\left(1 - \frac{1}{t^2+1}\right)$$

$$= 8\pi\left(\frac{t^2+1-1}{t^2+1}\right) = 8\pi\left(\frac{t^2}{t^2+1}\right).$$

2.7 Inverse Functions

3. Since there exists a horizontal line that intersects the graph of f in more than one point, f is not a one-to-one function.

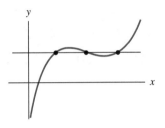

6. Since there exists a horizontal line that intersects the graph of f in more than one point, f is not a one-to-one function.

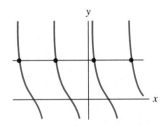

9. Since there exists a horizontal line that intersects the graph of f in more than one point, f is not a one-to-one function.

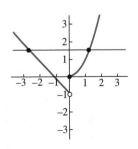

12. As in Example 2(b) in this section of the text, we need to find two values of x that have the same outputs. In this case we see that $f(2) = 0$ and $f(-1) = 0$, which shows that f is not one-to-one.

15. As in Example 2(b) in this section of the text, we assume that $f(x_1) = f(x_2)$. Then

$$\frac{2}{5x_1 + 8} = \frac{2}{5x_2 + 8} \quad \longleftarrow \quad \text{take reciprocals}$$

$$\frac{5x_1 + 8}{2} = \frac{5x_2 + 8}{2} \quad \longleftarrow \quad \text{multiply by 2}$$

$$5x_1 + 8 = 5x_2 + 8 \quad \longleftarrow \quad \text{subtract 8}$$

$$5x_1 = 5x_2 \quad \longleftarrow \quad \text{divide by 5}$$

$$x_1 = x_2$$

Since $f(x_1) = f(x_2)$ implies $x_1 = x_2$, we conclude that f is one-to-one.

18. As in Example 2(b) in this section of the text, we assume that $f(x_1) = f(x_2)$. Then

$$\frac{1}{x_1^3 + 1} = \frac{1}{x_2^3 + 1} \quad \longleftarrow \quad \text{take reciprocals}$$

$$x_1^3 + 1 = x_2^3 + 1 \quad \longleftarrow \quad \text{subtract 1}$$

$$x_1^3 = x_2^3 \quad \longleftarrow \quad \text{take cube roots}$$

$$x_1 = x_2$$

Since $f(x_1) = f(x_2)$ implies $x_1 = x_2$, we conclude that f is one-to-one.

21. We write the function as $y = 2/\sqrt{x}$. Solving for x, we have

$$y = \frac{2}{\sqrt{x}}$$

$$\sqrt{x} = \frac{2}{y}$$

$$x = \frac{4}{y^2}.$$

Relabeling the variables, we obtain $y = 4/x^2$ Thus, $f^{-1}(x) = 4/x^2$. The domain of f^{-1} is the range of f or $\{x \mid x > 0\}$ and the range of f^{-1} is the domain of f or $\{y \mid y > 0\}$.

24. We solve $y = -2x + 1$ for x:

$$y = -2x + 1$$

$$2x = 1 - y$$

$$x = \frac{1}{2}(1 - y).$$

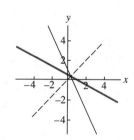

Relabeling the variables, we have $f^{-1}(x) = \frac{1}{2}(1 - x)$. The graphs of f and f^{-1} are shown with the graph of f^{-1} being blue.

27. We solve $y = 2 - \sqrt{x}$ for x:

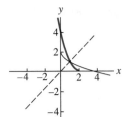

$$y = 2 - \sqrt{x}$$
$$\sqrt{x} = 2 - y$$
$$x = (2 - y)^2.$$

Relabeling the variables, we have $f^{-1}(x) = (2-x)^2 = (x-2)^2$. Since $y = (x-2)^2$ is a parabola, it is not one-to-one, and we must restrict the values of x. To do this, we can note from the graph of f^{-1} (shown in blue in the figure) that $x \le 2$. Alternatively, the range of f is $\{y \mid y \le 2\}$ so the domain of f^{-1} must be $\{x \mid x \le 2\}$. Hence, $f^{-1}(x) = (x-2)^2$, $x \le 2$.

30. The domain of f is $\{x \mid 5x + 8 \ne 0\} = \{x \mid x \ne \frac{8}{5}\}$. Then the range of f^{-1} is $\{y \mid y \ne \frac{8}{5}\}$. To find the domain of f^{-1} we find a formula for f^{-1} by solving $y = 2/(5x + 8)$ for x:

$$y = \frac{2}{5x + 8}$$

$$5x + 8 = \frac{2}{y}$$

$$5x = \frac{2}{y} - 8 = \frac{2 - 8y}{y}$$

$$x = \frac{2 - 8y}{5y}.$$

Thus,

$$f^{-1}(x) = \frac{2 - 8x}{5x}$$

and the domain of f^{-1} is $\{x \mid x \ne 0\}$. Hence, the range of f is $\{y \mid y \ne 0\}$.

33. When $x = 2$, $f(2) = 2(2)^3 + 2(2) = 2(8) + 4 = 20$, so the point on the graph of f is $(2, 20)$. Then, the corresponding point on the graph of f^{-1} is $(20, 2)$.

36. When $x = \frac{1}{2}$,

$$f\left(\frac{1}{2}\right) = \frac{4(1/2)}{1/2 + 1} = \frac{2}{3/2} = \frac{4}{3},$$

so the point on the graph of f is $(\frac{1}{2}, \frac{4}{3})$. Then, the corresponding point on the graph of f^{-1} is $(\frac{4}{3}, \frac{1}{2})$.

39. To graph f we use the fact that f is the inverse of f^{-1} and that the graph of the inverse of a function is the reflection of the graph of the original function reflected through the line $y = x$.

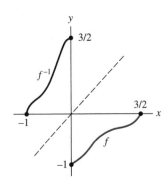

42. Solving for $y = (3 - 2x)^2$ for x we have

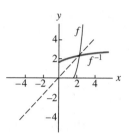

$$y = (3 - 2x)^2$$
$$-\sqrt{y} = 3 - 2x$$
$$2x = 3 + \sqrt{y}$$
$$x = \frac{1}{2}\left(3 + \sqrt{y}\right),$$

where we use the negative square root of y because, for $x \geq \frac{3}{2}$, $3 - 2x \leq 0$ and $\sqrt{y} \geq 0$ for all y. Relabeling the variables, we have $f^{-1}(x) = \frac{1}{2}\left(3 + \sqrt{x}\right)$, which has domain $\{x \mid x \geq 0\}$ or $[0, \infty)$. The graphs of f and f^{-1} are shown with the graph of f^{-1} in blue.

45. $f\left(f^{-1}(x)\right) = f\left(\frac{1}{5}x + 2\right) = 5\left(\frac{1}{5}x + 2\right) - 10 = x + 10 - 10 = x$

$f^{-1}\left(f(x)\right) = f^{-1}(5x - 10) = \frac{1}{5}(5x - 10) + 2 = x - 2 + 2 = x$

2.8 Translating Words into Functions

3. Let x and y be the nonnegative numbers. Then $x + y = 1$. Now, the sum of the square of x and twice the square of y is $x^2 + 2y^2$. From $x + y = 1$ we have $y = 1 - x$, so the function in this case is

$$f(x) = x^2 + 2(1 - x)^2 = x^2 + 2(1 - 2x + x^2) = 3x^2 - 4x + 2.$$

Since x and y are both nonnegative, we must have $0 \leq x \leq 1$ and $0 \leq y \leq 1$. (If, say $y > 1$, then we would have $x < 0$.) Thus

$$f(x) = 3x^2 - 4x + 2, \quad 0 \leq x \leq 1.$$

Alternatively, if we choose the independent variable of f to be y, we have $x = 1 - y$ and

$$f(y) = (1 - y)^2 + 2y^2 = 1 - 2y + y^2 + 2y^2 = 3y^2 - 2y + 1, \quad 0 \leq y \leq 1.$$

6. Let the sides of the rectangle be x and y as shown in the figure. Then $A = xy = 400$, $x, y > 0$. The perimeter of the rectangle is $P = 2x + 2y$. To express P in terms of just x, we use $y = 400/x$. Then

$$P(x) = 2x + 2\left(\frac{400}{x}\right) = 2x + \frac{800}{x} = \frac{2x^2 + 800}{x}, \quad x > 0.$$

Alternatively, from $xy = 400$ we have $x = 400/y$ and

$$P(x) = 2y + 2\left(\frac{400}{y}\right) = 2y + \frac{800}{y} = \frac{2y^2 + 800}{y}, \quad y > 0.$$

9. The distance between (x, y) and $(2, 3)$ is given by

$$d = \sqrt{(x-2)^2 + (y-3)^2}\,.$$

Since $x + y = 1$, we have $y = 1 - x$, so

$$d = \sqrt{(x-2)^2 + [(1-x) - 3]^2} = \sqrt{(x-2)^2 + (-x-2)^2}$$

$$= \sqrt{x^2 - 4x + 4 + x^2 + 4x + 4} = \sqrt{2x^2 + 8}\,.$$

The domain of d is $(-\infty, \infty)$.

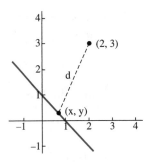

12. The diameter of a circle is twice its radius, that is, $d = 2r$ so $r = \frac{1}{2}d$. The area of a circle is then given by

$$A = \pi r^2 = \pi \left(\frac{1}{2}d\right)^2 = \frac{1}{4}\pi d^2.$$

15. Let the sides of the equilateral triangle each be of length x. Then, refering to the figure and using the Pythagorean theorem, we have

$$\left(\frac{x}{2}\right)^2 + h^2 = x^2$$

$$h^2 = x^2 - \frac{x^2}{4} = \frac{3}{4}x^2$$

$$x^2 = \frac{4}{3}h^2$$

$$x = \frac{2}{\sqrt{3}}h.$$

The area of the triangle is

$$A = \frac{1}{2}xh = \frac{1}{2}\left(\frac{2}{\sqrt{3}}h\right)h = \frac{1}{\sqrt{3}}h^2 = \frac{\sqrt{3}}{3}h^2.$$

(Whether or not you rationalize $1/\sqrt{3}$ as above is up to your instructor.) In this problem, $h > 0$ and the domain of the function is $(0, \infty)$.

18. As shown in the figure, we will bend the portion of the wire of length x into a square and the portion of length $L - x$ into a circle. Then the sides of the square are each $x/4$ and the circumference of the circle is $C = L - x$. Since the circumference of a circle is related to its radius by $C = 2\pi r$, we have $r = C/2\pi = (L - x)/2\pi$. The sum of the areas is

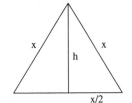

$$A = \text{area of square} + \text{area of circle}$$

$$= \left(\frac{x}{4}\right)^2 + \pi\left(\frac{L-x}{2\pi}\right)^2 = \frac{x^2}{16} + \pi\left(\frac{(L-x)^2}{4\pi^2}\right)$$

$$= \frac{x^2}{16} + \frac{(L-x)^2}{4\pi}\,.$$

In this problem, $0 < x < L$.

21. Let w be the width of the box, and h the height of the box. Then the length of the box is $3w$ and the volume of the box is

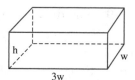

$$V = (3w)(w)(h) = 3w^2h = 450.$$

Since the box is open, its surface area is

$$S = 2wh + 2(3w)h + 3w(w) = 8wh + 3w^2.$$

From $3w^2h = 450$ we have $h = 150/w^2$, so

$$S = 8w\left(\frac{150}{w^2}\right) + 3w^2 = \frac{1200}{w} + 3w^2 = \frac{1200 + 3w^3}{w}.$$

In this problem, $w > 0$.

24. As shown in the figure, suppose the lower of the two airliners is traveling at 550 mi/h and the higher airliner is traveling at 500 mi/h. Then, after time t, the lower plane has traveled $550t$ miles and the higher plane has traveled $500t$ miles. From

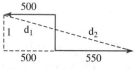

the figure, we see, using the Pythagorean Theorem, that the distance between the two planes is

$$d = d_1 + d_2 = \sqrt{(500t + 550t)^2 + 1^2} = \sqrt{(1050t)^2 + 1} = \sqrt{1{,}102{,}500t^2 + 1}.$$

In this problem, $t > 0$.

27. We want the maximum value of the difference of two numbers, x and x^2. The objective function is $D(x) = x - x^2$, where the domain of D is all real numbers.

30. We want to maximize the area as a function of the lengths of the sides of the plot of land. Let x be the side of the plot parallel to the two interior fences and let y be the length

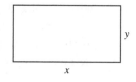

of the other side of each of the three subplots. See the figure. The total length of fencing used is $4x + 6y = 8000$. The area function is $A = x(3y) = 3xy$. Solving $4x + 6y = 8000$ for y, we have $6y = 8000 - 4x$ or $y = 4000/x - 2x/3$. Thus, the objective function is

$$A(x) = 3xy = 3x\left(\frac{4000}{3} - \frac{2}{3}x\right) = 4000x - 2x^2.$$

In this problem, $x > 0$.

33. We want to maximize area. The perimeter of the rectangle is $p = 2x + 2y$ and the area is $A = xy$. Solving $p = 2x + 2y$ for y, we obtain $2y = p - 2x$ or $y = \frac{1}{2}p - x$. Then, the objective function is

$$A(x) = xy = x\left(\frac{1}{2}p - x\right) = \frac{1}{2}px - x^2.$$

In this problem, since $A(x)$ must be positive, $0 \le x \le \frac{1}{2}p$.

36. The figure for Problem 35 in the text applies, except that the top of the box is closed. In this case we want to maximize volume. The total cost to construct each box is

$$2x^2 + xy + xy + xy + xy = 2x^2 + 4xy = 36.$$

The volume is $V = x^2 y$. Solving the cost equation for y, we obtain

$$2x^2 + 4xy = 36$$

$$4xy = 36 - 2x^2$$

$$y = \frac{36 - 2x^2}{4x} = \frac{18 - x^2}{2x}.$$

The objective function is

$$V(x) = x^2 y = x^2 \left(\frac{18 - x^2}{2x} \right) = 9x - \frac{1}{2} x^3.$$

In this problem, $x > 0$.

39. We want to minimize the area of the entire page, so let the width and height of the page be x and y, respectively, as shown in the figure. The area of the printed portion is $P = (x - 4)(y - 2) = 32$. The area of the entire page is $A = xy$. Solving $(x - 4)(y - 2) = 32$ for y, we obtain

$$(x - 4)(y - 2) = 32$$

$$y - 2 = \frac{32}{x - 4}$$

and thus,

$$y = \frac{32}{x - 4} + 2 = \frac{32 + 2x - 8}{x - 4} = \frac{24 + 2x}{x - 4}.$$

The objective function is

$$A(x) = xy = x \left(\frac{24 + 2x}{x - 4} \right) = \frac{24x + 2x^2}{x - 4}.$$

In this problem, $x > 4$, rather than $x > 0$, to allow for the side margins.

An alternative way to do the problem is to let the dimensions of the printed portion be x and y, where x is the height of the printed portion. This leads to $A(x) = 40 + 4x + 64/x$.

42. Figures for this problem are shown in the text. We want to minimize the surface area of the can. The volume is $V = \pi r^2 h = 32$. The surface area is

$$S = (\text{area of bottom}) + (\text{area of top}) + (\text{area of lateral side})$$

$$= \pi r^2 + \pi r^2 + 2\pi r h = 2\pi r^2 + 2\pi r h.$$

Solving $\pi r^2 h = 32$ for h, we have $h = 32/\pi r^2$. The objective function is

$$S(r) = 2\pi r^2 + 2\pi r h = 2\pi r^2 + 2\pi r \left(\frac{32}{\pi r^2} \right) = 2\pi r^2 + \frac{64\pi}{r}$$

$$= \frac{2\pi r^3 + 64\pi}{r}.$$

In this problem, we must have $r > 0$.

45. A picture of the trough is shown in the text and a figure for the end of the trough is shown to the right. We want to maximize the volume of the trough. The area of the end of the trough is the area of an isosceles triangle with sides 4, 4, and x. If we let h be the height of the triangle, then the area is $A = \frac{1}{2}xh$. To represent this strictly in terms of x, we use the Pythagorean theorem and solve for h:

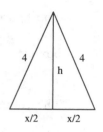

$$\left(\frac{x}{2}\right)^2 + h^2 = 4^2$$
$$\frac{x^2}{4} + h^2 = 16$$
$$x^2 + 4h^2 = 64$$
$$4h^2 = 64 - x^2$$
$$h^2 = \frac{64 - x^2}{4}$$
$$h = \frac{1}{2}\sqrt{64 - x^2}.$$

Thus, the area of the triangle is

$$A = \frac{1}{2}x\left(\frac{1}{2}\sqrt{64 - x^2}\right) = \frac{1}{4}x\sqrt{64 - x^2},$$

and the objective function is

$$S(x) = (\text{area of base}) \times (\text{height of trough}) = \left(\frac{1}{4}x\sqrt{64 - x^2}\right)(20) = 5x\sqrt{64 - x^2}.$$

In this problem, we must have $0 < x < 8$. (If $x \geq 8$, the three line segments will not form a triangle.)

2.9   Calculus Preview—The Tangent Line Problem

3. (a) We first compute $f(1 + h)$:

$$f(1 + h) = (1 + h)^2 - 3(1 + h)$$
$$= 1 + 2h + h^2 - 3 - 3h$$
$$= h^2 - h - 2.$$

Next, since $f(1) = 1^2 - 3(1) = 1 - 3 = -2$, we have

$$f(1 + h) - f(1) = h^2 - h - 2 - (-2) = h^2 - h = h(h - 1).$$

Finally,

$$\frac{f(1 + h) - f(1)}{h} = \frac{h(h - 1)}{h} = h - 1.$$

(b) $m_{\tan} = \lim\limits_{h \to 0} \dfrac{f(1+h) - f(1)}{h} = \lim\limits_{h \to 0} (h - 1) = -1.$

(c) The y-coordinate of the point of tangency is $f(1) = -2$, so using the point-slope form for the equation of a line we see that the tangent line is

$$y - (-2) = -1(x - 1) \qquad \text{or} \qquad y = -x - 1.$$

6. (a) We first compute $f(\frac{1}{2} + h)$:

$$f\left(\frac{1}{2} + h\right) = 8\left(\frac{1}{2} + h\right)^3 - 4$$

$$= 8\left(\frac{1}{8} + \frac{3}{4}h + \frac{3}{2}h^2 + h^3\right) - 4$$

$$= 1 + 6h + 12h^2 + h^3 - 4$$

$$= h^3 + 12h^2 + 6h - 3.$$

Next, since $f(\frac{1}{2}) = 8(\frac{1}{2})^3 - 4 = 8(\frac{1}{8}) - 4 = 1 - 4 = -3$, we have

$$f\left(\frac{1}{2} + h\right) - f\left(\frac{1}{2}\right) = h^3 + 12h^2 + 6h - 3 - (-3) = h(h^2 + 12h + 6).$$

Finally,

$$\frac{f(\frac{1}{2} + h) - f(\frac{1}{2})}{h} = \frac{h(h^2 + 12h + 6)}{h} = h^2 + 12h + 6.$$

(b) $m_{\tan} = \lim\limits_{h \to 0} \dfrac{f(\frac{1}{2} + h) - f(\frac{1}{2})}{h} = \lim\limits_{h \to 0} (h^2 + 12h + 6) = 6.$

(c) The y-coordinate of the point of tangency is $f(\frac{1}{2}) = -3$, so, using the point-slope form for the equation of a line, we see that the tangent line is

$$y - (-3) = 6\left(x - \frac{1}{2}\right) \qquad \text{or} \qquad y = 6x - 6.$$

9. (a) We first compute $f(4 + h)$:

$$f(4 + h) = \sqrt{4 + h}\,.$$

Next, since $f(4) = \sqrt{4} = 2$, we have

$$f(4 + h) - f(4) = \sqrt{4 + h} - 2.$$

Finally,

$$\frac{f(4+h) - f(4)}{h} = \frac{\sqrt{4+h} - 2}{h} = \frac{\sqrt{4+h} - 2}{h}\left(\frac{\sqrt{4+h} + 2}{\sqrt{4+h} + 2}\right)$$

$$= \frac{4 + h - 4}{h(\sqrt{4+h} + 2)} = \frac{h}{h(\sqrt{4+h} + 2)}$$

$$= \frac{1}{\sqrt{4+h} + 2}.$$

(b) $m_{\text{tan}} = \lim\limits_{h \to 0} \dfrac{f(4+h) - f(4)}{h} = \lim\limits_{h \to 0} \dfrac{1}{\sqrt{4+h}+2} = \dfrac{1}{2+2} = \dfrac{1}{4}$.

(c) The y-coordinate of the point of tangency is $f(4) = 2$ so, using the point-slope form for the equation of a line, we see that the tangent line is

$$y - 2 = \frac{1}{4}(x - 4) \qquad \text{or} \qquad y = \frac{1}{4}x + 1.$$

12. **(a)** We first find $f(x + h)$:

$$f(x + h) = -3(x + h) + 8 = -3x - 3h + 8.$$

Then

$$f(x + h) - f(x) = (-3x - 3h + 8) - (-3x + 8) = -3h,$$

and so

$$\frac{f(x + h) - f(x)}{h} = \frac{-3h}{h} = -3.$$

(b) The derivative of f is

$$f'(x) = \lim\limits_{h \to 0} \frac{f(x + h) - f(x)}{h} = \lim\limits_{h \to 0}(-3) = -3.$$

15. **(a)** We first find $f(x + h)$:

$$f(x + h) = 3(x + h)^2 - (x + h) + 7 = 3x^2 + 6xh + 3h^2 - x - h + 7.$$

Then

$$\begin{aligned} f(x + h) - f(x) &= (3x^2 + 6xh + 3h^2 - x - h + 7) - (3x^2 - x + 7) \\ &= 6xh + 3h^2 - h \\ &= h(6x + 3h - 1), \end{aligned}$$

and so

$$\frac{f(x + h) - f(x)}{h} = \frac{h(6x + 3h - 1)}{h} = 6x + 3h - 1.$$

(b) The derivative of f is

$$f'(x) = \lim\limits_{h \to 0} \frac{f(x + h) - f(x)}{h} = \lim\limits_{h \to 0}(6x + 3h - 1) = 6x - 1.$$

18. **(a)** We first find $f(x + h)$:

$$\begin{aligned} f(x + h) &= 2(x + h)^3 + (x + h)^2 \\ &= 2(x^3 + 3x^2h + 3xh^2 + h^3) + (x^2 + 2xh + h^2) \\ &= 2x^3 + 6x^2h + 6xh^2 + 2h^3 + x^2 + 2xh + h^2. \end{aligned}$$

Then

$$f(x+h) - f(x) = (2x^3 + 6x^2h + 6xh^2 + 2h^3 + x^2 + 2xh + h^2) - (2x^3 + x^2)$$
$$= 6x^2h + 6xh^2 + 2h^3 + 2xh + h^2$$
$$= h(6x^2 + 6xh + 2h^2 + 2x + h),$$

and so

$$\frac{f(x+h) - f(x)}{h} = \frac{h(6x^2 + 6xh + 2h^2 + 2x + h)}{h} = 6x^2 + 6xh + 2h^2 + 2x + h.$$

(b) The derivative of f is

$$f'(x) = \lim_{h \to 0} \frac{f(x+h) - f(x)}{h} = \lim_{h \to 0}(6x^2 + 6xh + 2h^2 + 2x + h) = 6x^2 + 2x.$$

21. (a) We first find $f(x+h)$:

$$f(x+h) = \frac{x+h}{x+h-1}.$$

Then

$$f(x+h) - f(x) = \frac{x+h}{x+h-1} - \frac{x}{x-1}$$
$$= \frac{(x+h)(x-1) - x(x+h-1)}{(x+h-1)(x-1)}$$
$$= \frac{x^2 + xh - x - h - x^2 - xh + x}{(x+h-1)(x-1)}$$
$$= -\frac{h}{(x+h-1)(x-1)},$$

and so

$$\frac{f(x+h) - f(x)}{h} = \frac{-\dfrac{h}{(x+h-1)(x-1)}}{h} = -\frac{h}{h(x+h-1)(x-1)}$$
$$= -\frac{1}{(x+h-1)(x-1)}.$$

(b) The derivative of f is

$$f'(x) = \lim_{h \to 0} \frac{f(x+h) - f(x)}{h} = \lim_{h \to 0}\left(-\frac{1}{(x+h-1)(x-1)}\right)$$
$$= -\frac{1}{(x-1)(x-1)} = -\frac{1}{(x-1)^2}.$$

24. (a) We first find $f(x+h)$:

$$f(x+h) = \frac{1}{(x+h)^2}.$$

Then

$$f(x+h) - f(x) = \frac{1}{(x+h)^2} - \frac{1}{x^2} = \frac{x^2 - (x+h)^2}{(x+h)^2 x^2}$$

$$= \frac{x^2 - x^2 - 2xh - h^2}{(x+h)^2 x^2} = -\frac{2xh - h^2}{(x+h)^2 x^2},$$

and so

$$\frac{f(x+h) - f(x)}{h} = \frac{-\dfrac{2xh - h^2}{(x+h)^2 x^2}}{h} = -\frac{2xh - h^2}{h(x+h)^2 x^2}$$

$$= -\frac{2x - h}{(x+h)^2 x^2}.$$

(b) The derivative of f is

$$f'(x) = \lim_{h \to 0} \frac{f(x+h) - f(x)}{h} = \lim_{h \to 0} \left(-\frac{2x - h}{(x+h)^2 x^2} \right)$$

$$= -\frac{2x}{x^2 x^2} = -\frac{2}{x^3}.$$

27. From Problem 15, $f'(x) = 6x - 1$, so when $x = 2$ the slope of the tangent line is

$$m_{\text{tan}} = f'(2) = 6(2) - 1 = 11.$$

The y-coordinate of the point on the graph of $y = f(x)$, when $x = 2$, is

$$f(2) = 3(2)^2 - 2 + 7 = 17,$$

so the point of tangency is $(2, 17)$. Using the point-slope form of the equation of a line, we then see that the tangent line is

$$y - 17 = 11(x - 2) \qquad \text{or} \qquad y = 11x - 5.$$

30. From Problem 18, $f'(x) = 6x^2 + 2x$, so at $x = -\frac{1}{2}$ the slope of the tangent line is

$$m_{\text{tan}} = f'\left(-\frac{1}{2}\right) = 6\left(-\frac{1}{2}\right)^2 + 2\left(-\frac{1}{2}\right) = \frac{3}{2} - 1 = \frac{1}{2}.$$

The y-coordinate of the point on the graph of $y = f(x)$ when $x = -\frac{1}{2}$ is

$$f\left(-\frac{1}{2}\right) = 2\left(-\frac{1}{2}\right)^3 + \left(-\frac{1}{2}\right)^2 = 2\left(-\frac{1}{8}\right) + \frac{1}{4} = 0.$$

Using the point-slope form of the equation of a line, we then see that the tangent line is

$$y - 0 = \frac{1}{2}\left[x - \left(-\frac{1}{2}\right)\right] \qquad \text{or} \qquad y = \frac{1}{2}x + \frac{1}{4}.$$

33. **(a)** We have $f(a) = 3a^2 + 1$ and

$$f(x) - f(a) = (3x^2 + 1) - (3a^2 + 1) = 3(x^2 - a^2) = 3(x + a)(x - a).$$

Then

$$\frac{f(x) - f(a)}{x - a} = \frac{3(x + a)(x - a)}{x - a} = 3(x + a).$$

(b) The derivative at $x = a$ is

$$f'(a) = \lim_{x \to a} \frac{f(x) - f(a)}{x - a} = \lim_{x \to a} 3(x + a) = 3(2a) = 6a.$$

36. **(a)** We have $f(a) = a^4$ and

$$f(x) - f(a) = x^4 - a^4 = (x - a)(x^3 + ax^2 + a^2x + a^3).$$

This result can be obtained by using synthetic division to factor $x - a$ out of $x^4 - a^4$. Now,

$$\frac{f(x) - f(a)}{x - a} = \frac{(x - a)(x^3 + ax^2 + a^2x + a^3)}{x - a} = x^3 + ax^2 + a^2x + a^3.$$

(b) The derivative at $x = a$ is

$$f'(a) = \lim_{x \to a} \frac{f(x) - f(a)}{x - a} = \lim_{x \to a} (x^3 + ax^2 + a^2x + a^3) = a^3 + a^3 + a^3 + a^3 = 4a^3.$$

39. **(a)** We have $f(a) = \sqrt{7a}$ and

$$f(x) - f(a) = \sqrt{7x} - \sqrt{7a} = \sqrt{7}\left(\sqrt{x} - \sqrt{a}\right).$$

Then

$$\frac{f(x) - f(a)}{x - a} = \frac{\sqrt{7}\left(\sqrt{x} - \sqrt{a}\right)}{x - a} = \frac{\sqrt{7}\left(\sqrt{x} - \sqrt{a}\right)}{x - a}\left(\frac{\sqrt{x} + \sqrt{a}}{\sqrt{x} + \sqrt{a}}\right)$$

$$= \frac{\sqrt{7}(x - a)}{(x - a)\left(\sqrt{x} + \sqrt{a}\right)} = \frac{\sqrt{7}}{\sqrt{x} + \sqrt{a}}.$$

(b) The derivative at $x = a$ is

$$f'(a) = \lim_{x \to a} \frac{f(x) - f(a)}{x - a} = \lim_{x \to a} \frac{\sqrt{7}}{\sqrt{x} + \sqrt{a}} = \frac{\sqrt{7}}{\sqrt{a} + \sqrt{a}} = \frac{\sqrt{7}}{2\sqrt{a}}.$$

Chapter 2 Review Exercises

3. We need $5 - x > 0$ (not $= 0$ because $\sqrt{5 - x}$ is in the denominator), so $x < 5$. The domain is $(-\infty, 5)$.

6. Symmetry with respect to the y-axis means $f(-x) = f(x)$.

9. The slope of the line through $(-2, 0)$ and $(0, -3)$ is

$$m = \frac{-3 - 0}{0 - (-2)} = -\frac{3}{2}.$$

12. The graph is a parabola opening downward, so we put the function in standard form to find the vertex:

$$\begin{aligned}
f(x) &= -x^2 + 6x - 21 \\
&= -(x^2 - 6x \quad\;) - 21 \\
&= -(x^2 - 6x + 9) - 21 + 9 \\
&= -(x - 3)^2 - 12.
\end{aligned}$$

The y-coordinate of the vertex is $(3, -12)$, so the range of the function is $(-\infty, -12]$.

15. The graph of $y = -5(x - 10)^2 + 2$ is the graph of $f(x) = x^2$ shifted 10 units to the right and 2 units up. Therefore, the vertex of the parabola $y = -5(x - 10)^2 + 2$ is $(10, 2)$.

18. To find the inverse function, we set $y = (x - 5)/(2x + 1)$ and solve for x:

$$\begin{aligned}
y &= \frac{x - 5}{2x + 1} \\
(2x + 1)y &= x - 5 \\
2xy + y &= x - 5 \\
2xy - x &= -y - 5 \\
x(2y - 1) &= -y - 5 \\
x &= -\frac{y + 5}{2y - 1}.
\end{aligned}$$

Interchanging x and y, we see that the inverse function is

$$f^{-1}(x) = -\frac{x + 5}{2x - 1}.$$

21. To see if the points are collinear, we find the slopes, m_1 and m_2, of the lines through $(0, 3), (2, 2)$, and $(2, 2), (6, 0)$, respectively:

$$m_1 = \frac{2 - 3}{2 - 0} = -\frac{1}{2} \quad \text{and} \quad m_2 = \frac{0 - 2}{6 - 2} = -\frac{2}{4} = -\frac{1}{2}.$$

Since the slopes are the same, the points are collinear. The statement is true.

24. If a nonzero function were to be symmetric with respect to the x-axis, its graph would fail the vertical line test for a function. The statement is true.

27. If a function were to take on the same value twice, say, $f(x_1) = f(x_2)$ for $x_1 \neq x_2$, it would not be one-to-one by definition. The statement is true.

30. The graphs of f and f^{-1} are symmetric with respect to the line $y = x$, so any point of intersection of the graphs must lie on this line. The statement is true.

33. If f is even, then $f(-x) = f(x)$ for all x, so f cannot be one-to-one (unless the domain of f is $\{0\}$, which is precluded by the fact that $a > 0$). The statement is true.

36. This is the definition of an increasing function as given in Section 2.3, so the statement is true.

39. Setting $x = 0$ in $y = 4 - 3f(x)$, we obtain $y(0) = 4 - 3f(0) = 4 - 3(1) = 1$, so the statement is true.

42. The most natural choice is to let $g(x) = x + 1$. Then $f(x) = 4x - \sqrt{x}$.

45. The domain of f^{-1} is the range of f, or $(-\pi/2, 3\pi/2)$. The range of f^{-1} is the domain of f, or $(-\infty, \infty)$.

48. Since

$$y(-x) = [\![-x]\!] + [\![-(-x)]\!] = [\![-x]\!] + [\![x]\!] = y(x),$$

we see that y is an even function. Next, if x is any nonnegative integer n, then

$$y(n) = [\![n]\!] + [\![-n]\!] = n - n = 0.$$

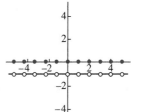

Finally, if $x = n + r$ for n, a nonnegative integer, and $0 < r < 1$, then

$$y(x) = [\![n + r]\!] + [\![-n - r]\!] = n + (n - 1) = -1.$$

The graph of y, shown in the figure, is discontinuous at every integer.

51. To find the inverse function, we set $y = (x + 1)^3$ and solve for x:

$$y = (x + 1)^3$$
$$y^{1/3} = x + 1$$
$$x = y^{1/3} - 1.$$

Interchanging x and y, we see that the inverse function is

$$f^{-1}(x) = x^{1/3} - 1.$$

54. If we set the origin at the vertex of the parabolic arch with the positive y-axis oriented upward, then the form of the equation of the arch is $y = -ax^2$, $-\frac{13}{2} < x < \frac{13}{2}$. To find a, we note that the point $\left(\frac{13}{2}, -8\right)$ is on the graph. That is,

$$-8 = -a\left(\frac{13}{2}\right)^2 \quad \text{or} \quad a = \frac{8}{169/4} = \frac{32}{169} \approx 0.18935.$$

The function is

$$y = -\frac{32}{169}x^2, \quad -\frac{13}{2} < x < \frac{13}{2}.$$

(Choosing a different location for the coordinate axis will change the answer.)

57. (a) See the figure in the text. Using distance = rate × time, we see that his distance from home base is $d(t) = 6t$. (Note that when $t = 15$ he is $d = 6(15) = 90$ feet from home base—that is, he is at first base.)

(b) If we let $t = 0$ when the player reaches second base, then his distance, d, from second base is also $d(t) = 6t$. Letting $D(t)$ represent the player's distance from home base at time t (measured from $t = 0$ at second base), we have, by the Pythagorean Theorem,

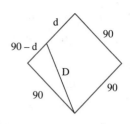

$$D^2 = 90^2 + (90 - d)^2$$
$$= 90^2 + 90^2 - 180d + d^2$$
$$= 16{,}200 - 180d + d^2$$
$$= 16{,}200 - 180(6t) + (6t)^2$$
$$= 16{,}200 - 1{,}080t + 36t^2$$
$$= 36(t^2 - 30t + 450).$$

Thus, his distance from home base at time t is

$$D(t) = 6\sqrt{t^2 - 30t + 450}.$$

60. We want to maximize area, so let x and h be the lengths of the walls, as shown in the figure. From the Pythagorean Theorem, we have

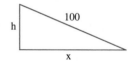

$$x^2 + h^2 = 100 \qquad \text{or} \qquad h = \sqrt{100 - x^2}.$$

Then, the area is

$$A(x) = \frac{1}{2}\, xh = \frac{1}{2}\, x\sqrt{100 - x^2}.$$

This is the objective function.

63. To find the tangent line, we first find the derivative of $f(x)$:

$$f(x + h) = -3(x + h)^2 + 16(x + h) + 12 = -3x^2 - 6xh - 3h^2 + 16x + 16h + 12,$$

$$f(x + h) - f(x) = (-3x^2 - 6xh - 3h^2 + 16x + 16h + 12) - (-3x^2 + 16x + 12)$$
$$= -6xh - 3h^2 + 16h,$$

$$\frac{f(x + h) - f(x)}{h} = \frac{-6xh - 3h^2 + 16h}{h} = -6x - 3h + 16,$$

$$f'(x) = \lim_{h \to 0} \frac{f(x + h) - f(x)}{h} = \lim_{h \to 0}(-6x - 3h + 16) = -6x + 16.$$

The slope of the tangent line when $x = 2$ is

$$m_{\text{tan}} = f'(2) = -6(2) + 16 = -12 + 16 = 4,$$

and the y-coordinate of the function when $x = 2$ is

$$f(2) = -3(2)^2 + 16(2) + 12 = 32.$$

Using the point-slope form of the equation of a line, we see that the tangent line is

$$y - 32 = 4(x - 2) \qquad \text{or} \qquad y = 4x + 24.$$

Chapter 3

Polynomial and Rational Functions

3.1 | Polynomial Functions

3. The graph of $y = (x-2)^3 + 2$ is the graph of $y = x^3$ shifted 2 units to the right and then shifted vertically upward 2 units. The graphs of $y = x^3$ (shown in blue) and $y = (x-2)^3 + 2$ are shown.

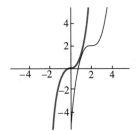

6. The graph of $y = x^4 - 1$ is the graph of $y = x^4$ shifted vertically downward 1 unit. The graphs of $y = x^4$ (shown in blue) and $y = x^4 - 1$ are shown.

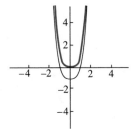

9. The function $f(x) = -2x^3 + 4x$ is odd because all powers of x are odd.

12. Since

$$f(x) = x^3(x+2)(x-2) = x^3(x^2 - 4) = x^5 - 4x^3,$$

we see that the function is odd because all powers of x are odd.

15. The graph corresponds to an odd function because of its end behavior. Since the graph is tangent to the x-axis at $x = 0$, but does not pass through the x-axis, $x = 0$ is a zero of even multiplicity. Thus, the graph corresponds to the function in (e).

18. The graph corresponds to an even function because of its end behavior. Since the graph is heading downward as $x \to -\infty$ and $x \to \infty$, the lead coefficient must be negative. Because the graph passes through the x-axis but is flattened at $(1, 0)$, the zero at $x = 1$ has odd multiplicity of at least 3. Thus, the graph corresponds to the function in (d).

21. The function is $f(x) = -x^3 + x^2 + 6x$.

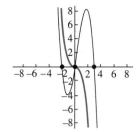

End Behavior: Ignoring all terms but the first, we see that the graph of f resembles the graph of $y = -x^3$ for large $|x|$. This graph is shown in blue in the figure.

Symmetry: Since both even and odd powers of x occur, the graph possesses no y-axis or origin symmetry.

Intercepts: Since $f(0) = 0$, the y-intercept is $(0,0)$. Solving

$$-x^3 + x^2 + 6x = -x(x^2 - x - 6) = -x(x - 3)(x + 2) = 0$$

we obtain x-intercepts $(-2,0), (0,0)$, and $(3,0)$.

The Graph: From left to right, the graph falls (see the black graph in the figure) from the second quadrant and then, because -2 is a simple zero, the graph passes straight through $(-2,0)$. The graph then turns upward toward $(0,0)$, which it passes straight through since 0 is a simple zero. Next, the graph turns downward toward $(3,0)$, which it passes straight through since 3 is a simple zero, and heads downward into the fourth quadrant.

24. The function is

$$f(x) = (2 - x)(x + 2)(x + 1) = -x^3 - x^2 + 4x + 4.$$

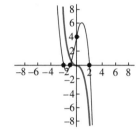

End Behavior: Ignoring all terms but the first, we see that the graph of f resembles the graph of $y = -x^3$ for large $|x|$. This graph is shown in blue in the figure.

Symmetry: Since both even and odd powers of x occur, the graph possesses no y-axis or origin symmetry.

Intercepts: Since $f(0) = 4$, the y-intercept is $(0,4)$. Solving

$$f(x) = (2 - x)(x + 2)(x + 1) = 0$$

we obtain x-intercepts $(-2,0), (-1,0)$, and $(2,0)$.

The Graph: From left to right, the graph falls (see the black graph in the figure) from the second quadrant and then, because -2 is a simple zero, the graph passes straight through $(-2,0)$. The graph then turns upward toward $(-1,0)$, which it passes straight through since -1 is a simple zero. Next, the graph continues upward through the y-intercept $(0,4)$ and then turns downward toward $(2,0)$. Since 2 is a simple zero, it passes straight through $(2,0)$ and continues downward into the fourth quadrant.

27. The function is

$$f(x) = (x^2 - x)(x^2 - 5x + 6) = x^4 - 6x^3 - x^2 - 6x.$$

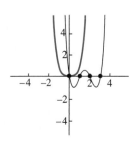

End Behavior: Ignoring all terms but the first, we see that the graph of f resembles the graph of $y = x^4$ for large $|x|$. This graph is shown in blue in the figure.

Symmetry: Since both even and odd powers of x occur, the graph possesses no y-axis or origin symmetry.

Intercepts: Since $f(0) = 0$, the y-intercept is $(0, 0)$. Solving

$$(x^2 - x)(x^2 - 5x + 6) = x(x - 1)(x - 2)(x - 3) = 0$$

we obtain x-intercepts $(0, 0), (1, 0), (2, 0)$, and $(3, 0)$.

The Graph: From left to right, the graph falls (see the black graph in the figure) from the second quadrant and then, because 0 is a simple zero, the graph passes straight through $(0, 0)$. The graph then turns upward toward $(1, 0)$, which it passes straight through since 1 is a simple zero. Next, the graph turns downward toward $(2, 0)$, which it passes straight through since 2 is a simple zero. Finally, it turns upward toward $(3, 0)$, passes through this point since 3 is a simple zero, and continues upward into the first quadrant.

30. The function is $f(x) = x^4 + 5x^2 - 6$.

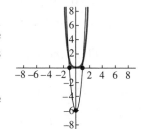

End Behavior: Ignoring all terms but the first, we see that the graph of f resembles the graph of $y = x^4$ for large $|x|$. This graph is shown in blue in the figure.

Symmetry: Since all powers of x (including $-6 = -6x^0$) are even, the graph is symmetric with respect to the y-axis.

Intercepts: Since $f(0) = -6$, the y-intercept is $(0, -6)$. Solving

$$x^4 + 5x^2 - 6 = (x^2 - 1)(x^2 + 6) = (x + 1)(x - 1)(x^2 + 6) = 0$$

we obtain x-intercepts $(-1, 0)$ and $(1, 0)$.

The Graph: From left to right, the graph falls (see the black graph in the figure) from the second quadrant and then, because -1 is a simple zero, the graph passes straight through $(-1, 0)$. Then the graph moves to $(0, -6)$. The graph is completed using symmetry with respect to the y-axis.

33. The function is $f(x) = x^4 + 3x^3$.

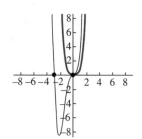

End Behavior: Ignoring all terms but the first, we see that the graph of f resembles the graph of $y = x^4$ for large $|x|$. This graph is shown in blue in the figure.

Symmetry: Since both even and odd powers of x occur, the graph possesses no y-axis or origin symmetry.

Intercepts: Since $f(0) = 0$, the y-intercept is $(0, 0)$. Solving

$$x^4 + 3x^3 = x^3(x + 3) = 0$$

we obtain x-intercepts $(-3, 0)$ and $(0, 0)$.

The Graph: From left to right, the graph falls (see the black graph in the figure) from the second quadrant and then, because -3 is a simple zero, the graph passes straight through $(-3, 0)$. The graph then turns upward toward $(0, 0)$, where it flattens out as it passes through the x-intercept, since 0 is a root of multiplicity 3. It then continues upward into the first quadrant.

36. The function is $f(x) = (x - 2)^5 - (x - 2)^3$, but we use the function $g(x) = x^5 - x^3$, which is the function $f(x)$ shifted 2 units to the right. We analyze $g(x)$.

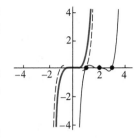

End Behavior: Ignoring all terms but the first, we see that the graph of f resembles the graph of $y = x^5$ for large $|x|$. This graph is shown in blue in the figure.

Symmetry: Since all powers of x in $g(x)$ are odd, the graph of $g(x)$ is symmetric with respect to the origin.

Intercepts: Since $g(0) = 0$, the y-intercept of $g(x)$ is $(0, 0)$. Solving

$$x^5 - x^3 = x^3(x^2 - 1) = x^3(x + 1)(x - 1) = 0$$

we obtain x-intercepts $(-1, 0), (0, 0),$ and $(1, 0)$ for $g(x)$.

The Graph: From left to right, the graph of $g(x)$ rises through the third quadrant (see the dashed graph in the figure) and then, because -1 is a simple zero, the graph passes straight through $(-1, 0)$. The graph then turns downward toward $(0, 0)$, where it flattens out and then passes through since 0 is a zero of multiplicity 3. The rest of the graph of $g(x)$ is completed using symmetry with respect to the origin. Finally, the graph of $g(x)$ (shown as a dashed curve in the figure) is shifted 2 units to the right to obtain the graph of $f(x)$, shown in black.

39. The function is $f(x) = -\frac{1}{2} x^2 (x + 2)^3 (x - 2)^2$.

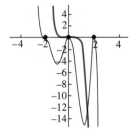

End Behavior: Ignoring all terms but the first, we see that the graph of f resembles the graph of $y = -x^7$ for large $|x|$. This graph is shown in blue in the figure.

Symmetry: Since $f(-1) = -\frac{9}{2}$ and $f(1) = -\frac{9}{2}$, there is a possibility that the function is even, but this cannot be the case for any seventh degree polynomial. Thus, the graph possesses no y-axis or origin symmetry.

Intercepts: Since $f(0) = 0$, the y-intercept is $(0, 0)$. Solving

$$-\frac{1}{2} x^2 (x + 2)^3 (x - 2)^2 = 0$$

we obtain x-intercepts $(-2, 0), (0, 0),$ and $(2, 0)$.

The Graph: From left to right, the graph falls (see the black graph in the figure) from the second quadrant and then, because -2 is a zero of multiplicity 3, it flattens out as it passes through $(-2, 0)$. The graph continues to fall into the third quadrant, but must at some point turn upward toward $(0, 0)$ where it is tangent to the x-axis and turns downward into the fourth quadrant. It must at some point turn back upward toward $(2, 0)$, where it is tangent to the x-axis and turns back downward into the fourth quadrant. (Tangency, without passing through the x-axis, occurs at both $(0, 0)$ and $(2, 0)$ because these are both zeros of even multiplicity 2.) The graph continues downward in the fourth quadrant as indicated by the blue graph of $y = -x^7$.

42. The function has simple zeros at $x = 1$ and $x = 3$. It has a zero of even multiplicity at $x = 4$. We want the lowest degree polynomial, so let the zero at $x = 4$ have multiplicity 2. The function then has the form $f(x) = a(x - 1)(x - 3)(x - 4)^2$ for some real number $a \neq 0$. Since $f(2) = 20$ we have

$$a(2 - 1)(2 - 3)(2 - 4)^2 = a(1)(-1)(4)^2 = -4a = 20,$$

so $a = -5$ and $f(x) = -5(x - 1)(x - 3)(x - 4)^2$.

45. We have

$$f(0) = 0^3 - 2(0)^2 + 14(0) - 3k = -3k = 10,$$

so $k = -\frac{10}{3}$.

48. (a) The graph crosses (as opposed to bounces off) the x-axis at the x-intercept $(5, 0)$ when the multiplicity of the zero at $x = 5$ is odd. Since $2m$, for m a positive integer, is never odd, there are no values of m for which the graph of f crosses the x-axis at $(5, 0)$.

    (b) The graph crosses (as opposed to bounces off) the x-axis at the x-intercept $(-1, 0)$ when the multiplicity of the zero at $x = -1$ is odd. Since $2n - 1$, for n a positive integer, is always odd, the graph of f crosses the x-axis at $(-1, 0)$ for all positive integer values of n.

51. From the figure of the piece of cardboard shown in the text, we see that $0 < x < 15$, because you cannot cut out more that half the width of the cardboard. Thus, the domain of $V(x)$ is all real numbers in the interval $(0, 15)$. The volume of a rectangular box is area of base × height, so, in this case,

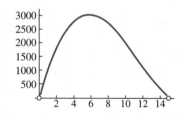

$$V(x) = (30 - 2x)(40 - 2x)x.$$

3.2   Division of Polynomial Functions

3. By long division,

$$
\begin{array}{r}
5x - 12 \\
x^2 + x - 1 \overline{\smash{\big)}\ 5x^3 - 7x^2 + 4x + 1} \\
\underline{5x^3 + 5x^2 - 5x} \\
-12x^2 + 9x + 1 \\
\underline{-12x^2 - 12x + 12} \\
21x - 11
\end{array}
$$

The quotient is $q(x) = 5x - 12$, the remainder is $r(x) = 21x - 1$, and
$$5x^3 - 7x^2 + 4x + 1 = (x^2 + x - 1)(5x - 12) + (21x - 11).$$

6. Using $(2x + 1)^2 = 4x^2 + 4x + 1$ and long division, we have

$$
\begin{array}{r}
\frac{1}{4}x \\
4x^2 + 4x + 1 \overline{\smash{\big)}\ x^3 + x^2 + x + 1} \\
\underline{x^3 + x^2 + \frac{1}{4}x } \\
\frac{3}{4}x + 1
\end{array}
$$

The quotient is $q(x) = \frac{1}{4}x$, the remainder is $r(x) = \frac{3}{4}x + 1$, and
$$x^3 + x^2 + x + 1 = (4x^2 + 4x + 1)\left(\frac{1}{4}x\right) + \left(\frac{3}{4}x + 1\right).$$

9. By long division,

$$
\begin{array}{r}
6x^2 + 4x + 1 \\
x^3 - 2 \overline{\smash{\big)}\ 6x^5 + 4x^4 + x^3 + 0x^2 + 0x + 0} \\
\underline{6x^5 + 0x^4 + 0x^3 - 12x^2 } \\
4x^4 + x^3 + 12x^2 + 0x + 0 \\
\underline{4x^4 + 0x^3 + 0x^2 - 8x } \\
x^3 + 12x^2 + 8x + 0 \\
\underline{x^3 + 0x^2 + 0x - 2} \\
12x^2 + 8x + 2
\end{array}
$$

The quotient is $q(x) = 6x^2 + 4x + 1$, the remainder is $r(x) = 12x^2 + 8x + 2$, and
$$6x^5 + 4x^4 + x^3 = (x^3 - 2)(6x^2 + 4x + 1) + (12x^2 + 8x + 2).$$

12. From the Remainder Theorem, the remainder r when $f(x) = 3x^2 + 7x - 1$ is divided by $x + 3$ is $f(-3)$. Thus,
$$r = f(-3) = 3(-3)^2 + 7(-3) - 1 = 5.$$

15. From the Remainder Theorem, the remainder r when $f(x) = x^4 - x^3 + 2x^2 + 3x - 5$ is divided by $x - 3$ is $f(3)$. Thus,
$$r = f(3) = 3^4 - 3^3 + 2(3)^2 + 3(3) - 5 = 76.$$

18. To find $f\left(\frac{1}{4}\right)$ we use the Remainder Theorem with dividend $f(x) = 6x^2 + 4x - 2$ and divisor $d(x) = x - \frac{1}{4}$:

$$
\begin{array}{r}
6x + \frac{11}{2} \\
x - \frac{1}{4} \overline{\smash{\big)}\ 6x^2 + 4\,x - 2} \\
\underline{6x^2 - \frac{3}{2}x } \\
\frac{11}{2}x - 2 \\
\underline{\frac{11}{2}x - \frac{11}{8}} \\
-\frac{5}{8}
\end{array}
$$

The remainder is $r = -\frac{5}{8}$, so $f\left(\frac{1}{4}\right) = r = -\frac{5}{8}$.

21. To find $f\left(\frac{1}{2}\right)$ we use the Remainder Theorem with dividend $f(x) = 3x^4 - 5x^2 + 20$ and divisor $d(x) = x - \frac{1}{2}$:

$$
\begin{array}{r}
3x^3 + \frac{3}{2}x^2 - \frac{17}{4}x - \frac{17}{8} \\
\hline
x - \frac{1}{2} \,) \; 3x^4 + 0x^3 - 5x^2 + \quad 0x + 20 \\
3x^4 - \frac{3}{2}x^3 \\
\hline
\frac{3}{2}x^3 - 5x^2 + \quad 0x + 20 \\
\frac{3}{2}x^3 - \frac{3}{4}x^2 \\
\hline
-\frac{17}{4}x^2 + \quad 0x + 20 \\
-\frac{17}{4}x^2 + \frac{17}{8}x \\
\hline
-\frac{17}{8}x + 20 \\
-\frac{17}{8}x + \frac{17}{16} \\
\hline
\frac{303}{16}
\end{array}
$$

The remainder is $r = \frac{303}{16}$, so $f\left(\frac{1}{2}\right) = r = \frac{303}{16}$.

24. Using synthetic division to divide $f(x) = 4x^2 - 8x + 6$ by $x - \frac{1}{2}$, we have

$$
\begin{array}{r|rrr}
\frac{1}{2} & 4 & -8 & 6 \\
 & & 2 & -3 \\
\hline
 & 4 & -6 & \underline{|\,3} = r\,.
\end{array}
$$

The quotient is $q(x) = 4x - 6$ and the remainder is $r = 3$.

27. Using synthetic division to divide $f(x) = x^4 + 16$ by $x - 2$, we have

$$
\begin{array}{r|rrrrr}
2 & 1 & 0 & 0 & 0 & 16 \\
 & & 2 & 4 & 8 & 16 \\
\hline
 & 1 & 2 & 4 & 8 & \underline{|\,32} = r\,.
\end{array}
$$

The quotient is $q(x) = x^3 + 2x^2 + 4x + 8$ and the remainder is $r = 32$.

30. Using synthetic division to divide $f(x) = 2x^6 + 3x^3 - 4x^2 - 1$ by $x + 1$, we have

$$
\begin{array}{r|rrrrrrr}
-1 & 2 & 0 & 0 & 3 & -4 & 0 & -1 \\
 & & -2 & 2 & -2 & -1 & 5 & -5 \\
\hline
 & 2 & -2 & 2 & 1 & -5 & 5 & \underline{|\,-6} = r\,.
\end{array}
$$

The quotient is $q(x) = 2x^5 - 2x^4 + 2x^3 + x^2 - 5x + 5$ and the remainder is $r = -6$.

33. From synthetic division

$$
\begin{array}{r|rrr}
-3 & 4 & -2 & 9 \\
 & & 12 & 42 \\
\hline
 & 4 & -14 & \underline{|\,51} = r
\end{array}
$$

we see that $f(-3) = r = 51$.

36. We express $f(x)$ using all the coefficients, including 0. Then, by synthetic division, we have

$$
\begin{array}{r|rrrrrr}
-2 & 3 & 0 & 0 & 1 & 0 & -16 \\
   &   & -6 & 12 & -24 & 46 & -92 \\
\hline
   & 3 & -6 & 12 & -23 & 46 & \boxed{-108} = r
\end{array}
$$

so $f(-2) = r = -108$.

39. We are looking for a value of k so that when $f(x) = x^4 + x^3 + 3x^2 + kx - 4$ is divided by $d(x) = x^2 - 1$, the remainder is 0. Using long division, we have

$$
\begin{array}{r}
x^2 + x + 4 \\
x^2 - 1 \overline{\smash{\big)}\ x^4 + x^3 + 3x^2 + kx - 4} \\
\underline{x^4 - x^2} \\
x^3 + 4x^2 + kx - 4 \\
\underline{x^3 - x} \\
4x^2 + (k+1)x - 4 \\
\underline{4x^2 - 4} \\
(k+1)x + 0
\end{array}
$$

The remainder is $r(x) = (k+1)x$, which we want to be 0, so choose $k = -1$. Then, when $k = -1$, $f(x) = x^4 + x^3 + 3x^2 - x - 4$ is divisible by $d(x) = x^2 - 1$.

42. We are looking for a value of k so that when $f(x) = x^3 + kx^2 - 2kx + 4$ is divided by $d(x) = x + 2$, the remainder is 0. Using synthetic division, we have

$$
\begin{array}{r|rrrr}
2 & 1 & k & -2k & 4 \\
  &   & -2 & -2k+4 & 8k-8 \\
\hline
  & 1 & k-2 & -4k+4 & \boxed{8k-4} = r\ .
\end{array}
$$

The remainder is $r = 8k - 4$, which we want to be 0, so choose $k = \frac{1}{2}$. Then, when $k = \frac{1}{2}$, $f(x) = x^3 + \frac{1}{2}x^2 - x + 4$ is divisible by $d(x) = x + 2$.

3.3   Zeros and Factors of Polynomial Functions

3. Since

$$f(5) = 5^3 - 6(5)^2 + 6(5) + 5 = 125 - 150 + 30 + 5 = 10 \neq 0,$$

we see that 5 is not a zero of $f(x)$.

Alternatively, we could have used synthetic division

$$
\begin{array}{r|rrrr}
5 & 1 & -6 & 6 & 5 \\
  &   & 5 & -5 & 5 \\
\hline
  & 1 & -1 & 1 & \boxed{10} = r
\end{array}
$$

to see that $f(5) = 10$. Since $f(5) \neq 0$, we conclude that 5 is not a zero of $f(x)$.

6. Since

$$f(-2) = (-2)^3 - 4(-2)^2 - 2(-2) + 20 = -8 - 16 + 4 + 20 = 0,$$

we see that -2 is a zero of $f(x)$.

Alternatively, we could have used synthetic division

$$
\begin{array}{r|rrrr}
-2 & 1 & -4 & -2 & 20 \\
   &   & -2 & 12 & -20 \\
\hline
   & 1 & -6 & 10 & \;\boxed{0} = r
\end{array}
$$

to see that $f(-2) = 0$. Since $f(-2) = 0$, we conclude that -2 is a zero of $f(x)$.

9. Since 1 is a zero of $f(x)$, we know that $x - 1$ is a factor of $f(x) = 9x^4 + 69x^3 - 29x^2 - 41x - 8$. Then, by synthetic division

$$
\begin{array}{r|rrrrr}
1 & 9 & 69 & -29 & -41 & -8 \\
  &   & 9 & 78 & 49 & 8 \\
\hline
  & 9 & 78 & 49 & 8   & \;\boxed{0} = r
\end{array}
$$

we see that

$$f(x) = (x - 1)q(x) = (x - 1)(9x^3 + 78x^2 + 49x + 8).$$

Since $-\frac{1}{3}$ is a zero of $f(x)$, we know that $x + \frac{1}{3}$ is a factor of $q(x) = 9x^3 + 78x^2 + 49x + 8$. Then, by synthetic division

$$
\begin{array}{r|rrrr}
-\frac{1}{3} & 9 & 78 & 49 & 8 \\
             &   & -3 & -25 & -8 \\
\hline
             & 9 & 75 & 24 & \;\boxed{0} = r
\end{array}
$$

we see that

$$f(x) = (x - 1)\left(x + \frac{1}{3}\right)(9x^2 + 75x + 24).$$

Since $9x^2 + 75x + 24$ is quadratic, and knowing that $x + \frac{1}{3}$ is a factor (since it was a zero of multiplicity 2 for $f(x)$), we easily factor

$$9x^2 + 75x + 24 = (3x + 1)(3x + 24) = 3(3x + 1)(x + 8).$$

Thus, the fourth zero of $f(x)$ is 8 and

$$f(x) = 3(x - 1)\left(x + \frac{1}{3}\right)(3x + 1)(x + 8) = (x - 1)(3x + 1)^2(x + 8).$$

12. Using synthetic division

$$\begin{array}{r|rrr} -\frac{1}{2} & 10 & -27 & 11 \\ & & -5 & 16 \\ \hline & 10 & -32 & \boxed{27} = r \end{array}$$

we see that the remainder when $f(x) = 10x^2 - 27x + 11$ is divided by $x + \frac{1}{2}$ is $r = 27 \neq 0$. Thus, $x = -\frac{1}{2}$ is not a zero of $f(x)$.

15. Using synthetic division

$$\begin{array}{r|rrrr} \frac{1}{3} & 3 & -3 & 8 & -2 \\ & & 1 & -\frac{2}{3} & \frac{22}{9} \\ \hline & 3 & -2 & \frac{22}{3} & \boxed{\frac{4}{9}} = r \end{array}$$

we see that the remainder when $f(x) = 3x^3 - 3x^2 + 8x - 2$ is divided by $x - \frac{1}{3}$ is $r = \frac{4}{9} \neq 0$. Thus, $x = \frac{1}{3}$ is not a zero of $f(x)$.

18. Since $x(3x - 1) = 3x^2 - x$ is not linear we use long division, as opposed to synthetic division:

$$
\begin{array}{r}
x^2 - 2x + 1 \\
3x^2 - x \overline{\smash{\big)}\ 3x^4 - 7x^3 + 5x^2 - x + 0} \\
\underline{3x^4 - 0x^3 } \\
-6x^3 + 5x^2 - x + 0 \\
\underline{-6x^3 + 2x^2 } \\
3x^2 - x + 0 \\
\underline{3x^2 - x + 0} \\
0
\end{array}
$$

Since the remainder is 0, $x(3x - 1)$ is a factor of $f(x) = 3x^4 - 7x^3 + 5x^2 - x$, so $x = 0$ and $x = \frac{1}{3}$ are zeros of $f(x)$. From the long division we find that the quotient is $q(x) = x^2 - 2x + 1 = (x - 1)^2$, so the other zero is $x = 1$ and it has multiplicity 2. The complete factorization of $f(x)$ is

$$f(x) = 3x^4 - 7x^3 + 5x^2 - x = x(3x - 1)(x - 1)^2.$$

21. Using $i^2 = -1$ and $i^3 = -i$ we have

$$f(2i) = 3(2i)^3 - 5(2i)^2 + 12(2i) - 20 = 3(8i^3) - 5(4i^2) + 24i - 20$$
$$= -24i + 20 + 24i - 20 = 0.$$

Thus, $x = 2i$ is a zero of $f(x)$. Since the coefficients of $f(x)$ are real numbers, another zero is $x = -2i$, and

$$(x - 2i)(x + 2i) = x^2 - 4i^2 = x^2 + 4$$

is a factor of $f(x)$. To find the remaining factors we use long division:

$$
\begin{array}{r}
3x - 5 \\
x^2 + 4 \overline{\smash{\big)}\ 3x^3 - 5x^2 + 12x - 20} \\
\underline{3x^3 + 0x^2 + 12x} \\
-5x^2 + 0x - 20 \\
\underline{-5x^2 + 0x - 20} \\
0
\end{array}
$$

Thus, the other factor of $f(x)$ is $3x - 5$, so the zeros of $f(x)$ are $2i$, $-2i$, and $\frac{5}{3}$, and the complete factorization of $f(x)$ is

$$f(x) = 3x^3 - 5x^2 + 12x - 20 = (x - 2i)(x + 2i)(3x - 5).$$

24. We use $i^2 = -1$, $i^3 = -i$, and $i^4 = 1$. Then

$$
\begin{aligned}
f(-i) &= 4(-i)^4 - 8(-i)^3 + 9(-i)^2 - 8(-i) + 5 \\
&= 4(1)^4 - 8(-i^3) + 9(i^2) + 8i + 5 \\
&= 4 + 8(-i) - 9 + 8i + 5 = 0,
\end{aligned}
$$

and we see that $x = -i$ is a zero of $f(x)$. Since the coefficients of $f(x)$ are real numbers, another zero is $x = i$, and

$$(x + i)(x - i) = x^2 - i^2 = x^2 + 1$$

is a factor of $f(x)$. To find the remaining factors we use long division:

$$
\begin{array}{r}
4x^2 - 8x + 5 \\
x^2 + 1 \overline{\smash{\big)}\ 4x^4 - 8x^3 + 9x^2 - 8x + 5} \\
\underline{4x^4 + 0x^3 + 4x^2} \\
-8x^3 + 5x^2 - 8x + 5 \\
\underline{-8x^3 + 0x^2 - 8x} \\
5x^2 + 0x + 5 \\
\underline{5x^2 + 0x + 5} \\
0
\end{array}
$$

Since $4x^2 - 8x + 5$ does not factor with integer coefficients, we use the quadratic formula to solve $4x^2 - 8x + 5 = 0$:

$$x = \frac{8 \pm \sqrt{(-8)^6 - 4(4)(5)}}{2(4)} = \frac{8 \pm \sqrt{64 - 80}}{8} = \frac{8 \pm \sqrt{-16}}{8} = 1 \pm \frac{1}{2}i.$$

The zeros of $f(x)$ are i, $-i$, $1 + \frac{1}{2}i$, and $1 - \frac{1}{2}i$, and the complete factorization of $f(x)$ is

$$
\begin{aligned}
f(x) &= (x + i)(x - i)\left[x - \left(1 + \frac{1}{2}i\right)\right]\left[x - \left(1 - \frac{1}{2}i\right)\right] \\
&= (x + i)(x - i)\left(x - 1 - \frac{1}{2}i\right)\left(x - 1 + \frac{1}{2}i\right).
\end{aligned}
$$

27. A fourth degree polynomial with the given zeros is

$$f(x) = (x-2)(x-1)(x+3)^2 = (x^2 - 3x + 2)(x^2 + 6x + 9)$$
$$= x^4 + 3x^3 - 7x^2 - 15x + 18.$$

30. Since $5i$ is a zero, so is $-5i$; and since $2 - 3i$ is a zero, so is $2 + 3i$. Thus, a fourth degree polynomial with these four zeros is

$$f(x) = (x - 5i)(x + 5i)[x - (2 - 3i)][x - (2 + 3i)]$$
$$= (x^2 - 25i^2)[(x - 2) + 3i][(x - 2) - 3i]$$
$$= (x^2 + 25)[(x - 2)^2 - 9i^2] = (x^2 + 25)(x^2 - 4x + 4 + 9)$$
$$= (x^2 + 25)(x^2 - 4x + 13) = x^4 - 4x^3 + 38x^2 - 100x + 325.$$

33. Setting $x = 0$, $4x - 5 = 0$, and $2x - 1 = 0$ we see that the zeros of $f(x)$ are $x = 0$ with multiplicity 1, $x = \frac{5}{4}$ with multiplicity 2, and $x = \frac{1}{2}$ with multiplicity 3.

36. Setting $x^2 + 25 = 0$ we have $x^2 = -25$, so $x = 5i$ and $x = -5i$ are zeros of multiplicity 1. Setting $x^2 - 5x + 4 = (x - 1)(x - 4)$ we see that $x = 1$ and $x = 4$ are zeros of multiplicity 2.

39. Since the graph is tangent to the x-axis at $x = -2$, the function has a zero of even multiplicity at $x = -2$. Since the graph passes through the x-axis at $x = 4$ without being tangent to the x-axis at that point, $x = 4$ is a simple zero. The degree of $f(x)$ is 3, so $f(x) = a(x + 2)^2(x - 4) = a(x^3 - 12x - 16)$. When $x = 0$, we see from the graph that $y = f(0) = 1$. From above, $f(0) = -16a$ so $a = -\frac{1}{16}$ and

$$f(x) = -\frac{1}{16}x^3 + \frac{3}{4}x + 1.$$

3.4 | Real Zeros of Polynomial Functions |

3. The factors of $a_0 = -3$ and $a_3 = 1$ are

$$p : \pm 1, \pm 3 \qquad \text{and} \qquad s : \pm 1,$$

respectively, so possible rational zeros are

$$\frac{p}{s} : \pm 1, \pm 3.$$

It is easily seen by direct substitution into $f(x)$ that $x = -1$, $x = 1$, and $x = -3$ are not zeros of $f(x)$. We then use synthetic division to test $x = 3$:

$$\underline{3|} \quad 1 \quad 0 \quad -8 \quad -3$$
$$\qquad \quad 3 \quad 9 \quad 3$$
$$\overline{\qquad 1 \quad 3 \quad 1 \quad |0 = r}.$$

Thus, $x = 3$ is a zero of $f(x)$ and

$$f(x) = (x - 3)(x^2 + 3x - 1).$$

Since $x^2 + 3x - 1$ does not factor with integer coefficients, the only rational zero of $f(x)$ is 3.

6. The factors of $a_0 = -2$ and $a_4 = 8$ are

$$p : \pm 1, \pm 2 \quad \text{and} \quad s : \pm 1, \pm 2, \pm 4, \pm 8,$$

respectively, so possible rational zeros are

$$\frac{p}{s} : \pm 1, \pm \frac{1}{2}, \pm \frac{1}{4}, \pm \frac{1}{8}, \pm 2.$$

(There are a number of other possible ratios, such as $\frac{2}{8}$, but they can all be reduced to one of the forms listed—for example, $\frac{2}{8} = \frac{1}{4}$.) As usual, it is just as easy to check -1 and 1 directly. In this case, $f(-1) = 27$ and $f(1) = 15$ so neither -1 nor 1 is a zero of $f(x)$. Using synthetic division, we find $f(-\frac{1}{2}) = \frac{9}{2}$, so $-\frac{1}{2}$ is not a zero. Next, we use synthetic division to compute $f(\frac{1}{2})$:

$$
\begin{array}{r|rrrrr}
\frac{1}{2} & 8 & -2 & 15 & -4 & -2 \\
 & & 4 & 1 & 8 & 2 \\
\hline
 & 8 & 2 & 16 & 4 & \boxed{0} = r \, .
\end{array}
$$

Thus, $x = \frac{1}{2}$ is a zero of $f(x)$ and

$$f(x) = \left(x - \frac{1}{2} \right) (8x^3 + 2x^2 + 16x + 4).$$

Since all coefficients of $q(x) = 8x^3 + 2x^2 + 16x + 4$ are positive, we need only look for negative zeros. We use synthetic division to compute $q(-\frac{1}{4})$:

$$
\begin{array}{r|rrrr}
-\frac{1}{4} & 8 & 2 & 16 & 4 \\
 & & -2 & 0 & -4 \\
\hline
 & 8 & 0 & 16 & \boxed{0} = r \, .
\end{array}
$$

Thus, $f(-\frac{1}{4}) = (\frac{1}{4} - \frac{1}{2})q(\frac{1}{4}) = 0$, $x = -\frac{1}{4}$ is a zero of $f(x)$, and

$$f(x) = \left(x - \frac{1}{2} \right) \left(x + \frac{1}{4} \right) (8x^2 + 16).$$

Since $8x^2 + 16$ has no real zeros, the rational zeros of $f(x)$ are $\frac{1}{2}$ and $-\frac{1}{4}$.

9. The factors of $a_0 = 3$ and $a_4 = 6$ are

$$p : \pm 1, \pm 3 \quad \text{and} \quad s : \pm 1, \pm 2, \pm 3, \pm 6,$$

respectively, so possible rational zeros are

$$\frac{p}{s} : \pm 1, \frac{1}{2}, \pm \frac{1}{3}, \pm \frac{1}{6}, \pm 3, \pm \frac{3}{2}.$$

(There are a number of other possible ratios, such as $\frac{3}{6}$, but they can all be reduced to one of the forms listed—for example, $\frac{3}{6} = \frac{1}{2}$.) As usual, it is just as easy to check -1 and 1 directly. In this case, $f(-1) = 20$ and $f(1) = -6$, so neither -1 nor 1

is a zero of $f(x)$. Using synthetic division, we find $f(-\frac{1}{2}) = \frac{15}{2}$, $f(\frac{1}{2}) = -\frac{7}{4}$, and $f(-\frac{1}{3}) = \frac{154}{27}$, so $-\frac{1}{2}$, $\frac{1}{2}$, and $-\frac{1}{3}$ are not zeros. We use synthetic division to compute $f(\frac{1}{3})$:

$$
\begin{array}{r|rrrrr}
\frac{1}{3} & 6 & -5 & -2 & -8 & 3 \\
  &   & 2 & -1 & -1 & -3 \\
\hline
  & 6 & -3 & -3 & -9 & \boxed{0} = r \,.
\end{array}
$$

Thus, $x = \frac{1}{3}$ is a zero of $f(x)$ and

$$
f(x) = \left(x - \frac{1}{3} \right) (6x^3 - 3x^2 - 3x - 9) = 3 \left(x - \frac{1}{3} \right) (2x^3 - x^2 - x - 3).
$$

Possible rational zeros of $g(x) = 2x^3 - x^2 - x - 3$ are ± 1, $\pm \frac{1}{2}$, ± 3, and $\pm \frac{3}{2}$. Note that we have eliminated $\pm \frac{1}{6}$ as possible zeros. Using synthetic division, we find $f(-3) = 630$, $f(3) = 312$, and $f(-\frac{3}{2}) = \frac{231}{4}$, so -3, 3, and $-\frac{3}{2}$ are not zeros. Finally, we use synthetic division to compute $f(\frac{3}{2})$:

$$
\begin{array}{r|rrrr}
\frac{3}{2} & 2 & -1 & -1 & -3 \\
  &   & 3 & 3 & 3 \\
\hline
  & 2 & 2 & 2 & \boxed{0} = r \,.
\end{array}
$$

Thus, $x = \frac{3}{2}$ is a zero of $f(x)$ and

$$
f(x) = 3 \left(x - \frac{1}{3} \right) \left(x - \frac{3}{2} \right) (2x^2 + 2x + 2) = 6 \left(x - \frac{1}{3} \right) \left(x - \frac{3}{2} \right) (x^2 + x + 1).
$$

We use the quadratic formula to find the zeros of $h(x) = x^2 + x + 1$:

$$
x = \frac{-1 \pm \sqrt{1^2 - 4(1)(1)}}{2(1)} = -\frac{1}{2} \pm \frac{1}{2}\sqrt{-3} = -\frac{1}{2} \pm \frac{1}{2}\sqrt{3}\, i.
$$

Thus, the only rational zeros of $f(x)$ are $\frac{1}{3}$ and $\frac{3}{2}$.

In Problems 3, 6, and 9 we have shown in some detail how to solve problems like those in numbers 12 through 30 below. The following solutions will be more terse.

---

12. We begin by noting that $x = 0$ is a zero of $f(x)$ and

$$
f(x) = x(x^4 - 2x - 12).
$$

Working with $q(x) = x^4 - 2x - 12$ we see that the possible rational zeros of $q(x)$ are ± 1, ± 2, ± 3, ± 4, ± 6, and ± 12. Using direct substitution or synthetic division, we find $q(-1) = -9$, $q(1) = -13$, $q(-2) = 8$, and $q(2) = 0$. Thus, 2 is a zero of $q(x)$ and hence of $f(x)$. We then find from synthetic division that

$$
f(x) = x(x^4 - 2x - 12) = x(x - 2)(x^3 + 2x^2 + 4x + 6).
$$

We now use synthetic division with $p(x) = x^3 + 2x^2 + 4x + 6$, which has possible rational zeros ± 1, ± 2, ± 3, and ± 6. We already know that -1, 1, and -2 cannot be zeros (since they would then also be zeros of $q(x)$). Using synthetic division, we find $p(2) = 30$, $p(-3) = -15$, $p(3) = 63$, $p(-6) = -162$, and $p(6) = 318$. Thus, the only rational zeros of $f(x)$ are 0 and 2.

15. Since $f(x)$ has noninteger coefficients, we begin by multiplying $f(x)$ by the least common denominator 4 of the coefficients. This gives

$$g(x) = 4f(x) = 4\left(\frac{1}{2}x^3 - \frac{9}{4}x^2 + \frac{17}{4}x - 3\right) = 2x^3 - 9x^2 + 17x - 12.$$

Since the zeros of $f(x)$ and $g(x)$ are the same, we consider the set of all possible zeros of $g(x)$:

$$\frac{p}{s}: \ \pm 1, \ \pm\frac{1}{2}, \ \pm 2, \ \pm 3, \ \pm\frac{3}{2}, \pm 4, \ \pm 6, \ \pm 12.$$

Testing these, we find $g(-1) = -40$, $g(1) = -2$, $g(-2) = -98$, $g(2) = 2$, $g(-3) = -198$, $g(3) = 12$, $g(-\frac{3}{2}) = -\frac{129}{2}$, and $g(\frac{3}{2}) = 0$. From synthetic division, we have

$$f(x) = \frac{1}{4}g(x) = \frac{1}{4}(2x^3 - 9x^2 + 17x - 12) = \frac{1}{4}\left(x - \frac{3}{2}\right)(2x^2 - 6x + 8)$$

$$= \frac{1}{2}\left(x - \frac{3}{2}\right)(x^2 - 3x + 4).$$

Since $x^2 - 3x + 4$ does not factor with integer coefficients, there are no more rational zeros of $f(x)$. Thus, the only rational zero is $\frac{3}{2}$.

18. Since $f(x)$ has noninteger coefficients, we begin by multiplying $f(x)$ by the least common denominator 12 of the coefficients. This gives

$$g(x) = 12f(x) = 12\left(\frac{3}{4}x^3 + \frac{9}{4}x^2 + \frac{5}{3}x + \frac{1}{3}\right) = 9x^3 + 27x^2 + 20x + 4.$$

Since the zeros of $f(x)$ and $g(x)$ are the same, we consider the set of all possible zeros of $g(x)$:

$$\frac{p}{s}: \ -1, \ -\frac{1}{3}, \ -\frac{1}{9}, \ -2, \ -\frac{2}{3}, \ -\frac{2}{9}, \ -4, \ -\frac{4}{3}, \ -\frac{4}{9}.$$

Note that we need only consider negative possible zeros because all coefficients of $g(x)$ are positive. Testing the possible zeros using synthetic division, we find $g(-1) = 2$ and $g(-\frac{1}{3}) = 0$. Since $-\frac{1}{3}$ is a zero, we use the result of the synthetic division to write

$$f(x) = \frac{1}{12}g(x) = \frac{1}{12}(9x^3 + 27x^2 + 20x + 4) = \frac{1}{12}\left(x + \frac{1}{3}\right)(9x^2 + 24x + 12)$$

$$= \frac{3}{12}\left(x + \frac{1}{3}\right)(3x^2 + 8x + 4) = \frac{1}{4}\left(x + \frac{1}{3}\right)(3x + 2)(x + 2).$$

We see that $-\frac{2}{3}$ and -2 are also zeros of $f(x)$. Thus, the zeros of $f(x)$ are -2, $-\frac{2}{3}$, and $-\frac{1}{3}$.

21. We look first for any rational zeros by using direct substitution into $f(x)$ or synthetic division to test the possibilities

$$\frac{p}{s} : \pm 1, \pm\frac{1}{2}, \pm\frac{1}{4}, \pm\frac{1}{8}, \pm 3, \pm\frac{3}{2}, \pm\frac{3}{4}, \pm\frac{3}{8}.$$

We find $f(-1) = 11$, $f(1) = 5$, $f(-\frac{1}{2}) = \frac{35}{4}$, $f(\frac{1}{2}) = -\frac{1}{4}$, $f(-\frac{1}{4}) = \frac{95}{16}$, $f(\frac{1}{4}) = \frac{11}{16}$, $f(-\frac{1}{8}) = \frac{71}{16}$, $f(\frac{1}{8}) = \frac{55}{32}$, $f(-3) = -135$, $f(3) = 231$, $f(-\frac{3}{2}) = \frac{15}{4}$, $f(\frac{3}{2}) = \frac{99}{4}$, $f(-\frac{3}{4}) = \frac{171}{16}$, $f(\frac{3}{4}) = \frac{15}{16}$, $f(-\frac{3}{8}) = \frac{237}{32}$, and (finally!) $f(\frac{3}{8}) = 0$.

This last result follows from the synthetic division

$$
\begin{array}{r|rrrr}
\frac{3}{8} & 8 & 5 & -11 & 3 \\
  &   & 3 & 3 & -3 \\
\hline
  & 8 & 8 & -8 & \;\boxed{0} = r\,,
\end{array}
$$

so

$$f(x) = \left(x - \frac{3}{8}\right)(8x^2 + 8x - 8) = (8x - 3)(x^2 + x - 1).$$

Using the quadratic formula to solve $x^2 + x - 1 = 0$, we find

$$x = \frac{-1 \pm \sqrt{1^2 - 4(1)(-1)}}{2(1)} = -\frac{1}{2} \pm \frac{1}{2}\sqrt{5}.$$

The zeros of $f(x)$ are $\frac{3}{8}$, $-\frac{1}{2} - \frac{1}{2}\sqrt{5}$, and $-\frac{1}{2} + \frac{1}{2}\sqrt{5}$, and a factorization is

$$f(x) = (8x - 3)\left[x - \left(-\frac{1}{2} - \frac{1}{2}\sqrt{5}\right)\right]\left[x - \left(-\frac{1}{2} + \frac{1}{2}\sqrt{5}\right)\right]$$

$$= (8x - 3)\left(x + \frac{1}{2} + \frac{1}{2}\sqrt{5}\right)\left(x + \frac{1}{2} - \frac{1}{2}\sqrt{5}\right).$$

24. We look first for any rational zeros by using direct substitution in $f(x)$ or synthetic division to test the possibilities

$$\frac{p}{s} : \pm 1, \pm 2, \pm 3, \pm 4, \text{ etc.}$$

We find $f(-1) = 100$, $f(1) = 144$, $f(-2) = 36$, $f(2) = 100$, and $f(-3) = 0$. This result follows from the synthetic division

$$
\begin{array}{r|rrrrr}
-3 & 1 & -2 & -23 & 24 & 144 \\
   &   & -3 & 15 & 24 & -144 \\
\hline
   & 1 & -5 & -8 & 48 & \;\boxed{0} = r\,,
\end{array}
$$

so

$$f(x) = (x + 3)(x^3 - 5x^2 - 8x + 48).$$

We continue by testing $q(x) = x^3 - 5x^2 - 8x + 48$ for rational zeros. We already know that ± 1 and ± 2 cannot be zeros, so we start with $x = -3$ (in case it is a multiple zero of $f(x)$). From

$$
\begin{array}{r|rrrr}
-3 & 1 & -5 & -8 & 48 \\
   &   & -3 & 24 & -48 \\
\hline
   & 1 & -8 & 16 & \boxed{0} = r
\end{array}
$$

we see that -3 is a zero of $q(x)$, and hence at least a double root of $f(x)$. We now have, by simple factoring, that

$$f(x) = (x+3)^2(x - 8x + 16) = (x+3)^2(x-4)^2.$$

The zeros of $f(x)$ are -3 (multiplicity 2) and 4 (multiplicity 2).

27. We begin by noting that $x = 0$ is a zero of $f(x)$ and

$$f(x) = 4x(x^4 - 2x^3 - 6x^2 + 10x - 3).$$

We now check

$$\frac{p}{s} : \ \pm 1, \ \pm 3$$

to see if there are any rational zeros of $q(x) = x^4 - 2x^3 - 6x^2 + 10x - 3$. Using synthetic division, we find $q(-1) = -16$, $q(1) = 0$, and

$$q(x) = (x-1)(x^3 - x^2 - 7x + 3).$$

We continue by testing $q_1(x) = x^3 - x^2 - 7x + 3$ for rational zeros. We already know that -1 cannot be a zero, so we start with $x = 1$ (in case it is a multiple zero of $q(x)$). We find $q_1(1) = 4$, $q_1(-3) = -12$, $q_1(3) = 0$, and

$$q_1(x) = (x-3)(x^2 + 2x - 1).$$

Since $x^2 + 2x - 1$ does not factor, we use the quadratic formula:

$$x = \frac{-2 \pm \sqrt{2^2 - 4(1)(-1)}}{2(1)} = -1 \pm \frac{1}{2}\sqrt{4 + 4} = -1 \pm \sqrt{2}.$$

The zeros of $f(x)$ are 0, 1, 3, $-1 - \sqrt{2}$ and $-1 + \sqrt{2}$, and

$$f(x) = 4x(x-1)(x-3)\left[x - (-1 - \sqrt{2})\right]\left[x - (-1 + \sqrt{2})\right]$$
$$= 4x(x-1)(x-3)(x+1+\sqrt{2})(x+1-\sqrt{2}).$$

30. We first look for rational zeros by using synthetic division to test the possibilities

$$\frac{p}{s} : \ \pm 1, \ \pm 2, \ \pm 8, \ \pm 16, \ \pm 32, \ \pm 64.$$

We find $f(-1) = -27$, $f(1) = -27$, $f(-2) = 0$, and

$$f(x) = (x+2)(x^5 - 2x^4 - 8x^3 + 16x^2 + 16x - 32).$$

We now test $q(x) = x^5 - 2x^4 - 8x^3 + 16x^2 + 16x - 32$ for rational zeros starting with -2 (in case it is a multiple zero of $f(x)$). In this case, -2 is a zero of $q(x)$ so it is a multiple zero of $f(x)$ and

$$f(x) = (x+2)^2(x^4 - 4x^3 + 16x - 16).$$

Next, we test $q_1(x) = x^4 - 4x^3 + 16x - 16$ for rational zeros starting again with $x = -2$. We find $q_1(-2) = 0$ and

$$f(x) = (x+2)^3(x^3 - 6x^2 + 12x - 8).$$

Testing $q_2(x) = x^3 - 6x^2 + 12x - 8$ for rational zeros starting again with $x = -2$ we find $q_2(-2) = -64$, $q_2(2) = 0$, and

$$f(x) = (x+2)^3(x-2)(x^2 - 4x + 4) = (x+2)^3(x-2)(x-2)^2 = (x+2)^3(x-2)^3.$$

The zeros of $f(x)$ are -2 and 2, each with multiplicity 3.

33. To find the real zeros of $f(x) = 2x^4 + 7x^3 - 8x^2 - 25x - 6$ (which are the same as the real solutions of $2x^4 + 7x^3 - 8x^2 - 25x - 6 = 0$), we first find any rational zeros. Using synthetic division to test the possibilities

$$\frac{p}{s}: \quad \pm 1, \ \pm\frac{1}{2}, \ \pm 2, \ \pm 3, \ \pm\frac{3}{2}, \ \pm 6$$

we find $f(-1) = 6$, $f(1) = -30$, $f(-\frac{1}{2}) = \frac{15}{4}$, $f(\frac{1}{2}) = -\frac{39}{2}$, $f(-2) = -12$, $f(2) = 0$, and

$$f(x) = (x-2)(2x^3 + 11x^2 + 14x + 3).$$

Next, we look for rational zeros of $q(x) = 2x^3 + 11x^2 + 14x + 3$ using synthetic division. We start with $x = 2$ (in case it is a multiple zero of $f(x)$) and find $q(2) = 91$, $q(-3) = 6$, $q(3) = 198$, $q(-\frac{3}{2}) = 0$, and

$$q(x) = \left(x + \frac{3}{2}\right)(2x^2 + 8x + 2) = 2\left(x + \frac{3}{2}\right)(x^2 + 4x + 1).$$

We use the quadratic formula to solve $x^2 + 4x + 1$:

$$x = \frac{-4 \pm \sqrt{4^2 - 4(1)(1)}}{2(1)} = -2 \pm \frac{1}{2}\sqrt{16 - 4} = -2 \pm \sqrt{3}.$$

The real solutions of the original equation are 2, $-\frac{3}{2}$, $-2 - \sqrt{3}$, and $-2 + \sqrt{3}$.

36. To find the real zeros of $f(x) = 8x^4 - 6x^3 - 7x^2 + 6x - 1$ (which are the same as the real solutions of $8x^4 - 6x^3 - 7x^2 + 6x - 1 = 0$), we first find any rational zeros. Using synthetic division to test the possibilities

$$\frac{p}{s}: \quad \pm 1, \ \pm\frac{1}{2}, \ \pm\frac{1}{4}, \ \pm\frac{1}{8}$$

we find $f(-1) = 0$ and

$$f(x) = (x+1)(8x^3 - 14x^2 + 7x - 1).$$

Next, we look for rational zeros of $q(x) = 8x^3 - 14x^2 + 7x - 1$ using synthetic division. We start with $x = -1$ (in case it is a multiple zero of $f(x)$) and find $q(-1) = -30$, $q(1) = 0$, and

$$q(x) = (x - 1)(8x^2 - 6x + 1) = (x - 1)(2x - 1)(4x - 1).$$

The real solutions of $8x^4 - 6x^3 - 7x^2 + 6x - 1 = 0$ are -1, 1, $\frac{1}{2}$, and $\frac{1}{4}$.

39. Since $f(0) = -6 < 0$ and $f(1) = 1 > 0$, by the Intermediate Value Theorem, the function has a zero between $x = 0$ and $x = 1$. Hoping that the zero is rational, we note that the possibilities are

$$\frac{p}{s}: \quad \pm 1, \ \pm\frac{1}{2}, \ \pm\frac{1}{4}, \ \pm 2, \ \pm 3, \ \pm\frac{3}{2}, \ \pm\frac{3}{4}, \ \pm 6.$$

We only need to test the numbers between 0 and 1: $\frac{1}{4}$, $\frac{1}{2}$, $\frac{3}{4}$. Using synthetic division, we find $f(\frac{1}{4}) = -\frac{25}{8}$, $f(\frac{1}{2}) = -\frac{5}{4}$, and $f(\frac{3}{4}) = 0$. Thus, the zero between 0 and 1 is $\frac{3}{4}$.

42. A cubic polynomial with the given zeros is

$$
\begin{aligned}
g(x) &= \left(x - \frac{1}{2}\right)\left[x - (1 + \sqrt{3})\right]\left[x - (1 - \sqrt{3})\right] \\
&= \left(x - \frac{1}{2}\right)\left[(x - 1) - \sqrt{3}\right]\left[(x - 1) + \sqrt{3}\right] \\
&= \left(x - \frac{1}{2}\right)\left[(x - 1)^2 + 3\right] = \left(x - \frac{1}{2}\right)(x^2 - 2x + 1 + 3) \\
&= \left(x - \frac{1}{2}\right)(x^2 - 2x + 4) = x^3 - \frac{5}{2}x^2 + 5x - 2.
\end{aligned}
$$

Any nonzero constant multiple of $g(x)$ has the same zeros, so let

$$f(x) = ag(x) = ax^3 - \frac{5}{2}ax^2 + 5ax - 2a.$$

We want $5a = 2$, so $a = \frac{2}{5}$ and

$$f(x) = \frac{2}{5}x^3 - x^2 + 2x - \frac{4}{5}.$$

3.5 | Rational Functions |

3. We first note that the numerator and denominator of $f(x) = 1/(x - 2)$ have no common factors.

Vertical Asymptotes: Setting $x - 2 = 0$ we see that $x = 2$ is a vertical asymptote.

Horizontal Asymptote: The degree of the numerator is less than the degree of the denominator, so $y = 0$ is a horizontal asymptote.

Intercepts: Since $f(0) = -\frac{1}{2}$, the y-intercept is $(0, -\frac{1}{2})$. The numerator is never 0, so there are no x-intercepts.

Graph: The graph is the graph of $y = 1/x$ shifted 2 units to the right.

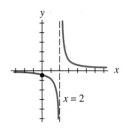

6. We first note that the numerator and denominator of $f(x) = x/(2x - 5)$ have no common factors.

   Vertical Asymptotes: Setting $2x - 5 = 0$ we see that $x = \frac{5}{2}$ is a vertical asymptote.

   Horizontal Asymptote: The degree of the numerator equals the degree of the denominator, so $y = \frac{1}{2}$ is the horizontal asymptote.

   Intercepts: Since $f(0) = 0$, the y-intercept is $(0, 0)$. Since the numerator is simply x, $(0, 0)$ is also the only x-intercept.

   Graph: The asymptotes are shown as dashed lines. The left branch has to be below the horizontal asymptote in order to pass through the intercept. The right branch has to be above the horizontal asymptote since there is no x-intercept to the right of $x = \frac{5}{2}$ and the graph does not cross $y = \frac{1}{2}$ since $x/(2x-5) = 1/2$ has no solution.

   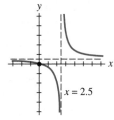

9. We first note that the numerator and denominator of $f(x) = (1 - x)/(x + 1)$ have no common factors.

   Vertical Asymptotes: Setting $x + 1 = 0$ we see that $x = -1$ is a vertical asymptote.

   Horizontal Asymptote: The degree of the numerator equals the degree of the denominator, so $y = -1/1 = -1$ is the horizontal asymptote.

   Intercepts: Since $f(0) = 1$, the y-intercept is $(0, 1)$. Setting $1 - x = 0$ we see that $x = 1$, or $(1, 0)$, is the x-intercept.

   Graph: Using long division, we find that

   $$\frac{1 - x}{x + 1} = -1 + \frac{2}{x + 1}.$$

   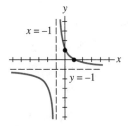

   Thus, the graph of $y = (1 - x)/(x + 1)$ is the graph of $y = 1/x$ stretched vertically by a factor of 2, shifted left 1 unit, and finally shifted 1 unit downward.

12. We first note that the numerator and denominator of $f(x) = 4/(x + 2)^3$ have no common factors.

    Vertical Asymptotes: Setting $(x+2)^3 = 0$ we see that $x = -2$ is a vertical asymptote.

    Horizontal Asymptote: The degree of the numerator is less than the degree of the denominator, so $y = 0$ is the horizontal asymptote.

    Intercepts: Since $f(0) = 4/2^3 = \frac{1}{2}$, the y-intercept is $(0, \frac{1}{2})$. The numerator is never 0, so there are no x-intercepts.

Graph: The left branch must lie entirely in the third quadrant since there are no x-intercepts, and $f(x) < 0$ for $x < -2$. The right branch must lie above the x-axis because it passes through $(0, \frac{1}{2})$ and there are no x-intercepts.

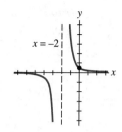

15. We first note that the numerator and denominator of $f(x) = x/(x^2 - 1)$ have no common factors.

Vertical Asymptotes: Setting $x^2 - 1 = (x + 1)(x - 1) = 0$ we see that $x = -1$ and $x = 1$ are vertical asymptotes.

Horizontal Asymptote: The degree of the numerator is less than the degree of the denominator, so $y = 0$ is the horizontal asymptote.

Intercepts: Since $f(0) = 0$, the y-intercept is $(0, 0)$. Since the numerator is simply x, $(0, 0)$ is also the only x-intercept.

Graph: We use the facts that the only x-intercept is $(0, 0)$ and the x-axis is a horizontal asymptote. For $x < -1$, $f(x) < 0$, so the left branch is below the x-axis. For $-1 < x < 0$, $f(x) > 0$, and for $0 < x < 1$, $f(x) < 0$, so the middle branch passes through the origin (as opposed to being tangent to the origin and lying strictly above or below the x-axis). For $x > 1$, $f(x) > 0$, so the right branch is above the x-axis.

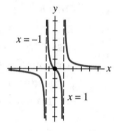

18. We write

$$f(x) = \frac{1}{x^2 - 2x - 8} = \frac{1}{(x + 2)(x - 4)}$$

and note that the numerator and denominator have no common factors.

Vertical Asymptotes: Setting $x^2 - 2x - 8 = (x + 2)(x - 4) = 0$ we see that $x = -2$ and $x = 4$ are vertical asymptotes.

Horizontal Asymptote: The degree of the numerator is less than the degree of the denominator, so $y = 0$ is the horizontal asymptote.

Intercepts: Since $f(0) = -\frac{1}{8}$, the y-intercept is $(0, -\frac{1}{8})$. The numerator is never 0, so there are no x-intercepts.

Graph: We use the facts that there are no x-intercepts and the x-axis is a horizontal asymptote. For $x < -2$, $f(x) > 0$, so the left branch is above the x-axis. For $-2 < x < 4$ the graph must lie entirely below the x-axis because it passes through $(0, -\frac{1}{8})$ and there are no x-intercepts. For $x > 4$, $f(x) > 0$, so the right branch is above the x-axis.

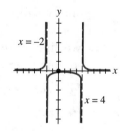

21. We write

$$f(x) = \frac{-2x^2 + 8}{(x-1)^2} = -\frac{2(x^2 - 4)}{(x-1)^2} = \frac{-2x^2 + 8}{x^2 - 2x + 1}$$

and note that the numerator and denominator have no common factors.

Vertical Asymptotes: Setting $(x-1)^2 = 0$ we see that $x = 1$ is a vertical asymptote.

Horizontal Asymptote: The degree of the numerator equals the degree of the denominator, so $y = -2/1 = -2$ is the horizontal asymptote.

Intercepts: Since $f(0) = 8$, the y-intercept is $(0, 8)$. Setting $x^2 - 4 = 0$ we see that $x = -2$ and $x = 2$ or $(-2, 0)$ and $(2, 0)$ are the x-intercepts.

Graph: Solving

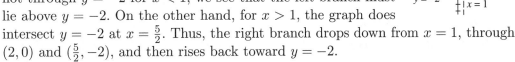

$$\frac{-2x^2 + 8}{x^2 - 2x + 1} = 2$$

we obtain $x = \frac{5}{2}$. Since the graph passes through $(-2, 0)$, but not through $y = -2$ for $x < 1$, we see that the left branch must lie above $y = -2$. On the other hand, for $x > 1$, the graph does intersect $y = -2$ at $x = \frac{5}{2}$. Thus, the right branch drops down from $x = 1$, through $(2, 0)$ and $(\frac{5}{2}, -2)$, and then rises back toward $y = -2$.

24. We write

$$f(x) = \frac{x^2 - 3x - 10}{x} = \frac{(x+2)(x-5)}{x}$$

and note that the numerator and denominator have no common factors.

Vertical Asymptotes: Setting the denominator equal to 0, we see that $x = 0$ or the y-axis is a vertical asymptote.

Slant Asymptote: Since the degree of the numerator is one greater than the degree of the denominator, the graph of $f(x)$ possesses a slant asymptote. From

$$f(x) = \frac{x^2 - 3x + 10}{x} = x - 3 + \frac{10}{x}$$

we see that $y = x - 3$ is a slant asymptote.

Intercepts: Since the y-axis is a vertical asymptote, the graph has no y-intercept. Setting $(x+2)(x-5) = 0$ we see that $x = -2$ and $x = 5$ or $(-2, 0)$ and $(5, 0)$ are the x-intercepts.

Graph: We can just about find the graph from the asymptotes and intercepts, but we need to determine if the graph crosses the slant asymptote. To do this we solve

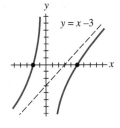

$$\frac{x^2 - 3x - 10}{x} = x - 3$$
$$x^2 - 3x - 10 = x^2 - 3x$$
$$-10 = 0.$$

Since there is no solution, the graph does not cross its slant asymptote.

27. We write

$$f(x) = \frac{x^2 - 2x - 3}{x - 1} = \frac{(x+1)(x-3)}{x - 1}$$

and note that the numerator and denominator have no common factors.

Vertical Asymptotes: Setting $x - 1 = 0$ we see that $x = 1$ is a vertical asymptote.

Slant Asymptote: Since the degree of the numerator is one greater than the degree of the denominator, the graph of $f(x)$ possesses a slant asymptote. Using synthetic division,

$$
\begin{array}{r|rrr}
\underline{1} & 1 & -2 & -3 \\
  &   & 1 & -1 \\
\hline
  & 1 & -1 & \underline{-4} = r
\end{array}
$$

we see that

$$f(x) = x - 1 - \frac{4}{x - 1}$$

and the slant asymptote is $y = x - 1$.

Intercepts: Since $f(0) = -3/(-1) = 3$, the y-intercept is $(0,3)$. Setting $(x + 1)(x - 3) = 0$, we see that $x = -1$ and $x = 3$ or $(-1,0)$ and $(3,0)$ are the x-intercepts.

Graph: We can just about find the graph from the asymptotes and intercepts, but we need to determine if the graph crosses the slant asymptote. To do this we solve

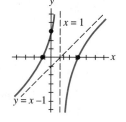

$$\frac{x^2 - 2x - 3}{x - 1} = x - 1$$
$$x^2 - 2x - 3 = x^2 - 2x + 1$$
$$-3 = 1.$$

Since there is no solution, the graph does not cross its slant asymptote.

30. We write

$$f(x) = \frac{5x(x+1)(x-4)}{x^2 + 1} = \frac{5x^3 - 15x^2 - 20x}{x^2 + 1}$$

and note that the numerator and denominator have no common factors.

Vertical Asymptotes: The denominator is never 0, so there are no vertical asymptotes.

Slant Asymptote: Since the degree of the numerator is one greater than the degree of the denominator, the graph of $f(x)$ possesses a slant asymptote. Using polynomial long division

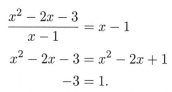

$$
\begin{array}{r}
5x - 15 \\
x^2 + 1 \overline{\smash{\big)}\, 5x^3 - 5x^2 - 20x + 0} \\
\underline{5x^3 + 0x^2 + 5x} \\
-15x^2 - 25x + 0 \\
\underline{-15x^2 + 0x - 15} \\
-25x + 15
\end{array}
$$

we see that

$$f(x) = 5x - 15 + \frac{-25x + 15}{x^2 + 1}$$

and the slant asymptote is $y = 5x - 15$.

Intercepts: Since $f(0) = 0$, the y-intercept is $(0,0)$. Setting $5x(x+1)(x-4) = 0$, we see that in addition to $(0,0)$, the other x-intercepts are $(-1,0)$ and $(4,0)$.

Graph: We first see if the graph of $f(x)$ intersects the slant asymptote:

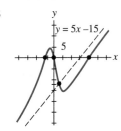

$$\frac{5x^3 - 15x^2 - 20x}{x^2 + 1} = 5x - 15$$
$$5x^3 - 15x^2 - 20x = 5x^3 - 15x^2 + 5x - 15$$
$$-20x = 5x - 15$$
$$15 = 25x$$
$$x = \frac{15}{25} = \frac{3}{5}.$$

Thus, the graph of $f(x)$ intersects the slant asymptote at $x = \frac{3}{5}$. Using the four points we have found and the slant asymptote, we obtain the graph of $f(x)$.

33. We write $f(x) = 4x(x-2)/(x-3)(x+4)$ and note that the numerator and denominator have no common factors.

Vertical Asymptotes: Setting $(x-3)(x+4) = 0$ we see that $x = 3$ and $x = -4$ are vertical asymptotes.

Horizontal Asymptote: The degree of the numerator equals the degree of the denominator, so $y = 4/1 = 4$ is the horizontal asymptote.

Intercepts: Since $f(0) = 0$, the y-intercept is $(0,0)$. Setting $4x(x-2) = 0$, we see that in addition to $(0,0)$, the other x-intercept is $(0,2)$.

Graph: To determine if the graph of $f(x)$ crosses the horizontal asymptote $y = 4$ we solve

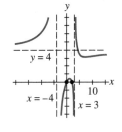

$$\frac{4x(x-2)}{(x-3)(x+4)} = 4$$
$$4x^2 - 8x = 4(x^2 + x - 12) = 4x^2 + 4x - 48$$
$$-8x = 4x - 48$$
$$48 = 12x$$
$$x = 4.$$

Since the graph of $f(x)$ does not intersect the horizontal asymptote $y = 4$ for $x < -4$ and there are no x-intercepts for $x < -4$, the left branch of the graph lies entirely above $y = 4$ and to the left of $x = -4$. The graph intersects the x-axis at $x = 0$ and $x = 2$ and does not intersect the horizontal asymptote $y = 4$ for $-4 < x < 3$, so it must lie mostly in quadrants III and IV, with a small portion in quadrant I. Finally, there are no x-intercepts for $x > 3$ and the graph intersects the horizontal asymptote $y = 4$ at $x = 4$, so the right branch must fall from just right of $x = 3$, through $(4,4)$ and then rise up again toward the horizontal asymptote $y = 4$.

36. For $f(x) = (x^3 + 2x - 4)/x^2$ the degree of the numerator is one greater than the degree of the denominator, so the graph has a slant asymptote. From a calculator or computer we find the graph shown to the right along with the slant asymptote.

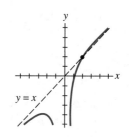

Slant Asymptote: Writing

$$f(x) = \frac{x^3 + 2x - 4}{x^2} = x + \frac{2x - 4}{x^2}$$

we see that the slant asymptote is $y = x$. To find where the graph of $f(x)$ intersects the line $y = x$ we solve

$$\frac{x^3 + 2x - 4}{x^2} = x$$
$$x^3 + 2x - 4 = x^3$$
$$2x - 4 = 0$$
$$x = 2.$$

Since the point is on the line $y = x$, we have $y = 2$, and the point of intersection is $(2, 2)$.

39. If the function is $f(x) = P(x)/Q(x)$ and the vertical asymptotes are $x = -1$ and $x = 2$, then we can take $Q(x) = (x + 1)(x - 2)$. For the horizontal asymptote to be $y = 3$ we want the numerator to be a second degree polynomial with leading coefficient 3. For example, we can use $P(x) = 3x^2$. But we also need an x-intercept to be $(3, 0)$, so we need $P(x)$ to have $x - 3$ as a factor. Thus, we use $P(x) = 3x(x - 3)$. The function

$$f(x) = \frac{P(x)}{Q(x)} = \frac{3x(x - 3)}{(x + 1)(x - 2)}$$

will satisfy the given conditions.

42. Writing

$$f(x) = \frac{x - 1}{(x + 1)(x - 1)}$$

we see that both the numerator and denominator have a factor of $x - 1$. Thus, there is a hole in the graph of $f(x)$ at $x = 1$ and we can write

$$f(x) = \frac{x - 1}{(x + 1)(x - 1)} = \frac{1}{x + 1}, \quad x \neq 1.$$

We see that the y-coordinate of the hole is $1/(1 + 1) = \frac{1}{2}$.

Intercepts: Since $f(0) = 1$, the y-intercept is $(0, 1)$. There are no x-intercepts because the numerator of $f(x)$ is never 0.

Graph: The graph of $f(x)$ is the graph of $y = 1/x$ shifted one unit left, with a hole at $(1, \frac{1}{2})$.

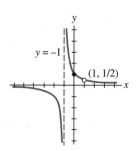

45. The vertical asymptote is $r = -5$ and the horizontal asymptote is $R = 5$. The origin is the only intercept. From the graph we see that the resistance R, as r becomes very large, approaches 5 from below.

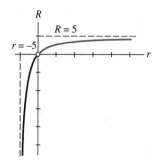

3.6   Partial Fractions

3. The partial fraction decomposition looks like

$$\frac{-9x + 27}{4x^2 - 4x - 5} = \frac{-9x + 27}{(x - 5)(x + 1)} = \frac{A}{x - 5} + \frac{B}{x + 1}.$$

Multiplying both sides of this equality by the denominator $(x - 5)(x + 1)$, we obtain

$$-9x + 27 = A(x + 1) + B(x - 5). \tag{3.1}$$

Since the denominator consists solely of linear factors, none of which is repeated, we can use the shortcut described on page 176 in the text. Letting $x = -1$ in (3.1) we have

$$-9(-1) + 27 = 36 = B(-1 - 5) = -6B,$$

so $B = -6$. Letting $x = 5$ in (3.1) we have

$$-9(5) + 27 = -18 = A(5 + 1) = 6A,$$

so $A = -3$. The partial fraction decomposition is

$$\frac{-9x + 27}{x^2 - 4x - 5} = -\frac{3}{x - 5} - \frac{6}{x + 1}.$$

6. The partial fraction decomposition looks like

$$\frac{1}{x(x - 2)(2x - 1)} = \frac{A}{x} + \frac{B}{x - 2} + \frac{C}{2x - 1}.$$

Multiplying both sides of this equality by the denominator $x(x - 2)(2x - 1)$, we obtain

$$1 = A(x - 2)(2x - 1) + Bx(2x - 1) + Cx(x - 2). \tag{3.2}$$

Since the denominator consists solely of linear factors, none of which is repeated, we can use the *Shortcut Worth Knowing* described on page 174 in the text. Letting $x = 0$ in (3.2) we have

$$1 = A(-2)(-1) = 2A,$$

so $A = \frac{1}{2}$. Letting $x = 2$ in (3.2) we have

$$1 = B(2)(4 - 1) = 6B,$$

so $B = \frac{1}{6}$. Letting $x = \frac{1}{2}$ in (3.2) we have

$$1 = C\left(\frac{1}{2}\right)\left(\frac{1}{2} - 2\right) = -\frac{3}{4}C,$$

so $C = -\frac{4}{3}$. The partial fraction decomposition is

$$\frac{1}{x(x - 2)(2x - 1)} = \frac{1/2}{x} + \frac{1/6}{x - 2} - \frac{4/3}{2x - 1}.$$

9. The partial fraction decomposition looks like

$$\frac{5x - 6}{(x - 3)^2} = \frac{A}{x - 3} + \frac{B}{(x - 3)^2}.$$

Multiplying both sides of this equality by the denominator $(x - 3)^2$ we obtain

$$5x - 6 = A(x - 3) + B \tag{3.3}$$
$$\text{or} \quad 5x - 6 = Ax + (-3A + B). \tag{3.4}$$

Letting $x = 3$ in (3.3) we have

$$5(3) - 6 = B,$$

so $B = 9$. Then (3.4) becomes

$$5x - 6 = Ax + (9 - 3A).$$

Equating coefficients of like terms, we see that $A = 5$. The partial fraction decomposition is

$$\frac{5x - 6}{(x - 3)^2} = \frac{5}{x - 3} + \frac{9}{(x - 3)^2}.$$

12. The partial fraction decomposition looks like

$$\frac{-4x + 6}{(x - 2)^2(x - 1)^2} = \frac{A}{x - 2} + \frac{B}{(x - 2)^2} + \frac{C}{x - 1} + \frac{D}{(x - 1)^2}.$$

Multiplying both sides of this equality by the denominator $(x-2)^2(x-1)^2$, we obtain

$$-4x + 6 = A(x - 2)(x - 1)^2 + B(x - 1)^2 + C(x - 2)^2(x - 1) + D(x - 2)^2 \tag{3.5}$$

or

$$-4x + 6 = (A + C)x^3 + (-4A + B - 5C + D)x^2 + (5A - 2B + 8C - 4D)x$$
$$+ (-2A + B - 4C + 4D). \tag{3.6}$$

Letting $x = 1$ in (3.5), we have

$$-4(1) + 6 = 2 = A(0) + B(0) + C(0) + D(-1)^2 = D,$$

so $D = 2$. Letting $x = 2$ in (3.5) we have

$$-4(2) + 6 = -2 = A(0) + B(1)^2 + C(0) + D(0) = B,$$

so $B = -2$, Substituting the values for B and D into (3.6) we have

$$-4x + 6 = (A + C)x^3 + (-4A - 5C)x^2 + (5A + 8C - 4)x + (-2A - 4C + 6).$$

Equating like coefficients, we get

$$\begin{array}{ccc} 5A + 8C - 4 = -4 & & 5A + 8C = 0 \\ & \text{or} & \\ -2A - 4C + 6 = 6 & & A + 2C = 0. \end{array}$$

Solving this system, we find $A = 0$ and $C = 0$. The partial fraction decomposition is

$$\frac{-4x + 6}{(x - 2)^2(x - 1)^2} = \frac{-2}{(x - 2)^2} + \frac{2}{(x - 1)^2}.$$

15. The partial fraction decomposition looks like

$$\frac{6x^2 - 7x + 11}{(x - 1)(x^2 + 9)} = \frac{A}{x - 1} + \frac{Bx + C}{x^2 + 9}.$$

Multiplying both sides of this equality by the denominator $(x - 1)(x^2 + 9)$ we obtain

$$6x^2 - 7x + 11 = A(x^2 + 9) + (Bx + C)(x - 1) \tag{3.7}$$
$$\text{or} \qquad 6x^2 - 7x + 11 = (A + B)x^2 + (-B + C)x + (9A - C). \tag{3.8}$$

Letting $x = 1$ in (3.7) we have

$$6(1)^2 - 7(1) + 11 = 10 = A(1 + 9) + [B(1) + C](0) = 10A,$$

so $A = 1$. Substituting $A = 1$ into (3.8) we have

$$6x^2 - 7x + 11 = (B + 1)x^2 + (-B + C)x + (9 - C).$$

Equating like coefficients, we get

$$B + 1 = 6 \qquad \text{so} \qquad B = 5$$

and

$$9 - C = 11 \qquad \text{so} \qquad C = -2.$$

The partial fraction decomposition is

$$\frac{6x^2 - 7x + 11}{(x - 1)(x^2 + 9)} = \frac{1}{x - 1} + \frac{5x - 2}{x^2 + 9}.$$

18. The partial fraction decomposition looks like

$$\frac{2x^2 - x + 7}{(x - 6)(x^2 + x + 5)} = \frac{A}{x - 6} + \frac{Bx + C}{x^2 + x - 5}.$$

Multiplying both sides of this equality by the denominator $(x - 6)(x^2 + x + 5)$, we obtain

$$2x^2 - x + 7 = A(x^2 + x + 5) + (Bx + C)(x - 6) \tag{3.9}$$

or $\quad 2x^2 - x + 7 = (A + B)x^2 + (A - 6B + C)x + (5A - 6C). \tag{3.10}$

Letting $x = 6$ in (3.9) we have

$$2(6)^2 - 6 + 7 = 73 = A(6^2 + 6 + 5) + (6B + C)(0) = 47A,$$

so $A = 73/47$. Substituting $A = 73/47$ into (3.10) we have

$$2x^2 - x + 7 = \left(\frac{73}{47} + B\right)x^2 + \left(\frac{73}{47} - 6B + C\right)x + \left(5 \cdot \frac{73}{47} - 6C\right).$$

Equating like coefficients, we get

$$\frac{73}{47} + B = 2 \quad \text{so} \quad B = 2 - \frac{73}{47} = \frac{21}{47}$$

and

$$5 \cdot \frac{73}{47} - 6C = 7 \quad \text{so} \quad C = \frac{1}{6}\left(\frac{365}{47} - 7\right) = \frac{1}{6}\left(\frac{36}{47}\right) = \frac{6}{47}.$$

The partial fraction decomposition is

$$\frac{2x^2 - x + 7}{(x - 6)(x^2 + x + 5)} = \frac{73/47}{x - 6} + \frac{(21/47)x + 6/47}{x^2 + x + 5}.$$

21. The partial fraction decomposition looks like

$$\frac{x^3}{(x^2 + 2)(x^2 + 1)} = \frac{Ax + B}{x^2 + 2} + \frac{Cx + D}{x^2 + 1}.$$

Multiplying both sides of this equality by the denominator $(x^2+2)(x^2+1)$, we obtain

$$x^3 = (Ax + B)(x^2 + 1) + (Cx + D)(x^2 + 2)$$

or $\quad x^3 = (A + C)x^3 + (B + D)x^2 + (A + 2C)x + (B + 2D). \tag{3.11}$

Equating like coefficients in (3.11) we get

$$A + C = 1 \tag{3.12}$$
$$B + D = 0 \tag{3.13}$$
$$A + 2C = 0 \tag{3.14}$$
$$B + 2D = 0. \tag{3.15}$$

Subtracting the equation in (3.12) from the equation in (3.14) we get

$$(A + 2C) - (A + C) = C = 0 - 1 = -1,$$

so $C = -1$. Substituting this value into (3.12) we find $A = 2$. Next, subtracting the equation in (3.13) from the equation in (3.15) we get $D = 0$. Substituting this value into (3.13) we find $B = 0$. Thus, the partial fraction decomposition is

$$\frac{x^3}{(x^2 + 2)(x^2 + 1)} = \frac{2x}{(x^2 + 2)} + \frac{-x}{x^2 + 1}.$$

24. The partial fraction decomposition looks like

$$\frac{2x^2}{(x - 2)(x^2 + 4)^2} = \frac{A}{x - 2} + \frac{Bx + C}{x^2 + 4} + \frac{Dx + E}{(x^2 + 4)^2}.$$

Multiplying both sides of this equality by the denominator $(x - 2)(x^2 + 4)^2$ we obtain

$$2x^2 = A(x^2 + 4)^2 + (Bx + C)(x - 2)(x^2 + 4) + (Dx + E)(x - 2) \qquad (3.16)$$

or

$$2x^2 = (A + B)x^4 + (-2B + C)x^3 + (8A + 4B - 2C + D)x^2$$
$$+ (-8B + 4C - 2D + E)x + (16A - 8C - 2E). \quad (3.17)$$

Letting $x = 2$ in (3.16) we have

$$2(2)^2 = 8 = 64A + (2B + C)(0)(8) + (2D + E)(0) = 64A,$$

so $A = \frac{1}{8}$. Substituting $A = \frac{1}{8}$ into (3.17), we get

$$2x^2 = \left(B + \frac{1}{8}\right)x^4 + (-2B + C)x^3 + (4B - 2C + D + 1)x^2 +$$
$$(-8B + 4C - 2D + E)x + (-8C - 2E + 2).$$

Equating like coefficients, we get

$$B + \frac{1}{8} = 0 \quad \text{so} \quad B = -\frac{1}{8}$$

$$-2B + C = 0 \quad \text{so} \quad C = 2B = 2\left(-\frac{1}{8}\right) = -\frac{1}{4}$$

$$4B - 8C + D + 1 = 2 \quad \text{so} \quad D = 1 - 4B + 2C = 1 + \frac{1}{2} - \frac{1}{2} = 1$$

$$-8B + 4C - 2D + E = 0 \quad \text{so} \quad E = 8B - 4C + 2D = -1 + 1 + 2 = 2.$$

Thus, the partial fraction decomposition is

$$\frac{2x^2}{(x - 2)(x^2 + 4)^2} = \frac{1/8}{x - 2} + \frac{(-1/8)x - 1/4}{x^2 + 4} + \frac{x + 2}{(x^2 + 4)^2}.$$

27. Using polynomial long division, we have

$$
2x^2 + 5x + 2 \overline{\smash{\big)}\ x^2 - 4x + 1} \quad \tfrac{1}{2}
$$
$$
\underline{x^2 + \tfrac{5}{2}x + 1}
$$
$$
-\tfrac{13}{2}x + 0
$$

so

$$
\frac{x^2 - 4x + 1}{2x^2 + 5x + 2} = \frac{1}{2} + \frac{(-13/2)x}{2x^2 + 5x + 2}.
$$

The partial fraction decomposition of the proper fraction looks like

$$
\frac{(-13/2)x}{2x^2 + 5x + 2} = \frac{(-13/2)x}{(x+2)(2x+1)} = \frac{A}{x+2} + \frac{B}{2x+1}.
$$

Multiplying both sides of this equality by the denominator $(x+2)(2x+1)$ we obtain

$$
-\frac{13}{2}x = A(2x+1) + B(x+2). \tag{3.18}
$$

Since the denominator consists solely of linear factors, none of which is repeated, we can use the *Shortcut Worth Knowing* described on page 174 in the text. Letting $x = -\tfrac{1}{2}$ in (3.18), we have

$$
-\frac{13}{2}\left(-\frac{1}{2}\right) = \frac{13}{4} = A(0) + B\left(-\frac{1}{2} + 2\right) = \frac{3}{2}B,
$$

so $B = \frac{13}{6}$. Letting $x = -2$ in (3.18), we have

$$
-\frac{13}{2}(-2) = 13 = A[2(-2)+1] + B(0) = -3A,
$$

so $A = -\frac{13}{3}$. The partial fraction decomposition is

$$
\frac{x^2 - 4x + 1}{2x^2 + 5x + 2} = \frac{1}{2} + \frac{-13/3}{x+2} + \frac{13/6}{2x+1}.
$$

30. Using polynomial long division, we have

$$
x^3 + 3x^2 + 3x + 1 \overline{\smash{\big)}\ x^3 + x^2 - x + 1} \quad 1
$$
$$
\underline{x^3 + 3x^2 + 3x + 1}
$$
$$
-2x^2 - 4x + 0
$$

so

$$
\frac{x^3 + x^2 - x + 1}{x^3 + 3x^2 + 3x + 1} = 1 + \frac{-2x^2 - 4x}{(x+1)^3}.
$$

The partial fraction decomposition of the proper fraction looks like

$$\frac{-2x^2 - 4x}{(x+1)^3} = \frac{A}{x+1} + \frac{B}{(x+1)^2} + \frac{C}{(x+1)^3}.$$

Multiplying both sides of this equality by the denominator $(x+1)^3$ we obtain

$$-2x^2 - 4x = A(x+1)^2 + B(x+1) + C \tag{3.19}$$
$$\text{or}\qquad -2x^2 - 4x = Ax^2 + (2A + B)x + (A + B + C). \tag{3.20}$$

Letting $x = -1$ in (3.19), we find

$$-2(-1)^2 - 4(-1) = 2 = A(0) + B(0) + C,$$

so $C = 2$. Equation (3.20) then becomes

$$-2x^2 - 4x = Ax^2 + (2A + B)x + (A + B + 2).$$

Equating coefficients of like terms, we have $A = -2$, $A + B + 2 = 0$, and $-2 + B + 2 = B = 0$. Thus, the partial fraction decomposition of the improper fraction is

$$\frac{x^3 + x^2 - x + 1}{x^3 + 3x^2 + 3x + 1} = 1 + \frac{-2}{x+1} + \frac{2}{(x+1)^3}.$$

3.7   The Area Problem

3. The width of each subinterval is $\Delta x = \frac{1-0}{4} = \frac{1}{4}$. The midpoints of the subintervals are $x_1^* = \frac{1}{8}$, $x_2^* = \frac{3}{8}$, $x_3^* = \frac{5}{8}$, and $x_4^* = \frac{7}{8}$. Then

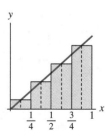

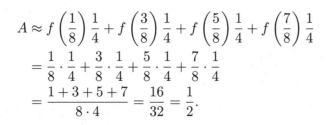

$$A \approx f\left(\frac{1}{8}\right)\frac{1}{4} + f\left(\frac{3}{8}\right)\frac{1}{4} + f\left(\frac{5}{8}\right)\frac{1}{4} + f\left(\frac{7}{8}\right)\frac{1}{4}$$
$$= \frac{1}{8}\cdot\frac{1}{4} + \frac{3}{8}\cdot\frac{1}{4} + \frac{5}{8}\cdot\frac{1}{4} + \frac{7}{8}\cdot\frac{1}{4}$$
$$= \frac{1+3+5+7}{8\cdot 4} = \frac{16}{32} = \frac{1}{2}.$$

6. By dividing $[-1, 2]$ into twelve subintervals, the width of each subinterval is $\Delta = \frac{2-(-1)}{12} = \frac{3}{12} = \frac{1}{4}$.

(a) The left-hand endpoints are $x_1^* = -1$, $x_2^* = -\frac{3}{4}$, $x_3^* = -\frac{1}{2}$, $x_4^* = -\frac{1}{4}$, $x_5^* = 0$, $x_6^* = \frac{1}{4}$, $x_7^* = \frac{1}{2}$, $x_8^* = \frac{3}{4}$, $x_9^* = 1$, $x_{10}^* = \frac{5}{4}$, $x_{11}^* = \frac{3}{2}$, and $x_{12}^* = \frac{7}{4}$. Then

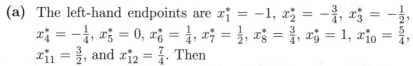

$$A \approx f(-1)\frac{1}{4} + f\left(-\frac{3}{4}\right)\frac{1}{4} + f\left(-\frac{1}{2}\right)\frac{1}{4} + f\left(-\frac{1}{4}\right)\frac{1}{4} + f(0)\frac{1}{4} + f\left(\frac{1}{4}\right)\frac{1}{4}$$

$$+ f\left(\frac{1}{2}\right)\frac{1}{4} + f\left(\frac{3}{4}\right)\frac{1}{4} + f(1)\frac{1}{4} + f\left(\frac{5}{4}\right)\frac{1}{4} + f\left(\frac{3}{2}\right)\frac{1}{4} + f\left(\frac{7}{4}\right)\frac{1}{4}$$

$$= 1 \cdot \frac{1}{4} + \frac{5}{4} \cdot \frac{1}{4} + \frac{3}{2} \cdot \frac{1}{4} + \frac{7}{4} \cdot \frac{1}{4} + 2 \cdot \frac{1}{4} + \frac{9}{4} \cdot \frac{1}{4} + \frac{5}{2} \cdot \frac{1}{4} + \frac{11}{4} \cdot \frac{1}{4} + 3 \cdot \frac{1}{4}$$

$$+ \frac{13}{4} \cdot \frac{1}{4} + \frac{7}{2} \cdot \frac{1}{4} + \frac{15}{4} \cdot \frac{1}{4}$$

$$= \frac{1}{4} + \frac{5}{16} + \frac{3}{8} + \frac{7}{16} + \frac{1}{2} + \frac{9}{16} + \frac{5}{8} + \frac{11}{16} + \frac{3}{4} + \frac{13}{16} + \frac{7}{8} + \frac{15}{16} = \frac{57}{8}.$$

(b) The right-hand endpoints are $x_1^* = -\frac{3}{4}$, $x_2^* = -\frac{1}{2}$, $x_3^* = -\frac{1}{4}$, $x_4^* = 0$, $x_5^* = \frac{1}{4}$, $x_6^* = \frac{1}{2}$, $x_7^* = \frac{3}{4}$, $x_8^* = 1$, $x_9^* = \frac{5}{4}$, $x_{10}^* = \frac{3}{2}$, $x_{11}^* = \frac{7}{4}$, and $x_{12}^* = 2$. Then

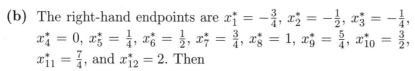

$$A \approx f\left(-\frac{3}{4}\right)\frac{1}{4} + f\left(-\frac{1}{2}\right)\frac{1}{4} + f\left(-\frac{1}{4}\right)\frac{1}{4} + f(0)\frac{1}{4} + f\left(\frac{1}{4}\right)\frac{1}{4} + f\left(\frac{1}{2}\right)\frac{1}{4}$$

$$+ f\left(\frac{3}{4}\right)\frac{1}{4} + f(1)\frac{1}{4} + f\left(\frac{5}{4}\right)\frac{1}{4} + f\left(\frac{3}{2}\right)\frac{1}{4} + f\left(\frac{7}{4}\right)\frac{1}{4} + f(2)\frac{1}{4}$$

$$= \frac{5}{4} \cdot \frac{1}{4} + \frac{3}{2} \cdot \frac{1}{4} + \frac{7}{4} \cdot \frac{1}{4} + 2 \cdot \frac{1}{4} + \frac{9}{4} \cdot \frac{1}{4} + \frac{5}{2} \cdot \frac{1}{4} + \frac{11}{4} \cdot \frac{1}{4} + 3 \cdot \frac{1}{4} + \frac{13}{4} \cdot \frac{1}{4}$$

$$+ \frac{7}{2} \cdot \frac{1}{4} + \frac{15}{4} \cdot \frac{1}{4} + 2 \cdot \frac{1}{4}$$

$$= \frac{5}{16} + \frac{3}{8} + \frac{7}{16} + \frac{1}{2} + \frac{9}{16} + \frac{5}{8} + \frac{11}{16} + \frac{3}{4} + \frac{13}{16} + \frac{7}{8} + \frac{15}{16} + \frac{1}{2} = \frac{63}{8}.$$

9. By dividing $[0, 5]$ into 5 subintervals, the width of each subinterval is $\Delta x = \frac{5-0}{5} = 1$. The midpoints of the subintervals are $x_1^* = \frac{1}{2}$, $x_2^* = \frac{3}{2}$, $x_3^* = \frac{5}{2}$, $x_4^* = \frac{7}{2}$, $x_5^* = \frac{9}{2}$. Then

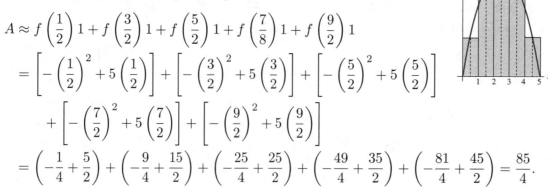

$$A \approx f\left(\frac{1}{2}\right)1 + f\left(\frac{3}{2}\right)1 + f\left(\frac{5}{2}\right)1 + f\left(\frac{7}{8}\right)1 + f\left(\frac{9}{2}\right)1$$

$$= \left[-\left(\frac{1}{2}\right)^2 + 5\left(\frac{1}{2}\right)\right] + \left[-\left(\frac{3}{2}\right)^2 + 5\left(\frac{3}{2}\right)\right] + \left[-\left(\frac{5}{2}\right)^2 + 5\left(\frac{5}{2}\right)\right]$$

$$+ \left[-\left(\frac{7}{2}\right)^2 + 5\left(\frac{7}{2}\right)\right] + \left[-\left(\frac{9}{2}\right)^2 + 5\left(\frac{9}{2}\right)\right]$$

$$= \left(-\frac{1}{4} + \frac{5}{2}\right) + \left(-\frac{9}{4} + \frac{15}{2}\right) + \left(-\frac{25}{4} + \frac{25}{2}\right) + \left(-\frac{49}{4} + \frac{35}{2}\right) + \left(-\frac{81}{4} + \frac{45}{2}\right) = \frac{85}{4}.$$

12. First, we divide $[0, 9]$ into 9 subintervals, each of width 1. The left-hand endpoints are $x_1^* = 0$, $x_2^* = 1$, $x_3^* = 2$, $x_4^* = 3$, $x_5^* = 4$, $x_6^* = 5$, $x_7^* = 6$, $x_8^* = 7$, and $x_9^* = 8$. The approximate area in this case is

$$A_L = f(0)1 + f(1)1 + f(2)1 + f(3)1 + f(4)1 + f(5)1 + f(6)1 + f(7)1 + f(8)1$$
$$= 0 + 3 + 1 + 3 + 5 + 3 + 7 + 4 + 6 = 32.$$

Second, we again divide $[0, 9]$ into 9 subintervals, each of width 1. The right-hand endpoints are $x_1^* = 1$, $x_2^* = 2$, $x_3^* = 3$, $x_4^* = 4$, $x_5^* = 5$, $x_6^* = 6$, $x_7^* = 7$, $x_8^* = 8$, and $x_9^* = 9$. The approximate area in this case is

$$A_R = f(1)1 + f(2)1 + f(3)1 + f(4)1 + f(5)1 + f(6)1 + f(7)1 + f(8)1 + f(9)1$$
$$= 3 + 1 + 3 + 5 + 3 + 7 + 4 + 6 + 3 = 35.$$

Chapter 3 Review Exercises

3. Writing

$$f(x) = x^2(x + 3)(x - 5) = x^4 - 2x^3 - 15x^2,$$

we see that the end behavior of the graph of $f(x)$ resembles the graph of the power function $f(x) = x^4$.

6. The y-intercept of the graph of $f(x)$ is the point $(0, f(0))$, or, in this case, $(0, \frac{8}{4})$ or $(0, 2)$.

9. Since

$$f(x) = \frac{x^3 - x}{4 - 2x^3},$$

we see that $y = 1/(-2) = -\frac{1}{2}$ is a horizontal asymptote.

12. Since the graph of a polynomial function of degree n can have at most $n - 1$ turning points, the graph of $f(x) = 3x^5 - 4x^2 + 5x - 2$ can have at most 4 turning points.

15. This is true; rational functions can have graphs with holes.

18. This is true, because if a, b, c, d, and x are all positive, then so is $f(x) = ax^3 + bx^2 + cx + d$ and $f(x)$ cannot be 0.

21. This is true because the powers of x with nonzero coefficients, namely 6 and 2, are both even.

24. Computing

$$f\left(\frac{1}{3}\right) = \frac{2(\frac{1}{3}) + 4}{3 - \frac{1}{3}} = \frac{2(\frac{1}{3}) + 4}{3 - \frac{1}{3}}\left(\frac{3}{3}\right) = \frac{2 + 12}{9 - 1} = \frac{14}{8} = \frac{7}{4}$$

we see that $(\frac{1}{3}, \frac{7}{4})$ is on the graph of $f(x)$. Thus, the statement is true.

27. This is false. The polynomial $f(x) = x^2 + 1$ has no rational zeros.

30. By long division,

$$
\begin{array}{r}
3x - \frac{2}{5} \\
5x^3 + x + 2 \overline{\smash{\big)}\ 15x^4 - 2x^3 + 0x^2 + 8x + 6} \\
\underline{15x^4 + 0x^3 + 3x^2 + 6x} \\
-2x^3 - 3x^2 + 2x + 6 \\
\underline{-2x^3 + 0x^2 - \frac{2}{5}x + \frac{4}{5}} \\
-3x^2 + \frac{12}{5}x + \frac{26}{5}
\end{array}
$$

Thus,

$$
\frac{15x^4 - 2x^3 + 8x + 6}{5x^3 + x + 2} = 3x - \frac{2}{5} + \frac{-3x^2 + \frac{12}{5}x + \frac{26}{5}}{5x^3 + x + 2}.
$$

33. By the Remainder Theorem, the remainder when the polynomial $f(x)$ is divided by $d(x) = x + 3 = x - (-3)$ is

$$
f(-3) = 5(-3)^3 - 4(-3)^2 + 6(-3) - 9 = -198.
$$

36. By the Remainder Theorem, the remainder when the polynomial $f(x)$ is divided by $d(x) = x - 1$ is

$$
f(1) = 36(1)^{98} - 40(1)^{25} + 18(1)^4 - 3(1)^7 + 40(1)^4 + 5(1)^2 - 1 + 2
$$
$$
= 36 - 40 + 18 - 3 + 40 + 5 - 1 + 2 = 57.
$$

39. Since

$$
f(2) = (2 - 3)^3 + 1 = (-1)^3 + 1 = -1 + 1 = 0,
$$

2 is a zero of $f(x)$. Writing

$$
f(x) = (x - 3)^3 + 1 = x^3 - 9x^2 + 27x - 27 + 1 = x^3 - 9x^2 + 27x - 26
$$

and using synthetic division to divide $f(x)$ by $x - 2$

$$
\begin{array}{r|rrrr}
2 & 1 & -9 & 27 & -26 \\
  &   & 2 & -14 & 26 \\
\hline
  & 1 & -7 & 13 & \boxed{0} = r,
\end{array}
$$

we see that $f(x) = (x - 2)(x^2 - 7x + 13)$. Since $x^2 - 7x + 13$ does not factor, we use the quadratic formula to solve $x^2 - 7x + 13 = 0$:

$$
x = \frac{-(-7) \pm \sqrt{(-7)^2 - 4(1)(13)}}{2(1)} = \frac{7}{2} \pm \frac{1}{2}\sqrt{-3} = \frac{7}{2} \pm \frac{1}{2}\sqrt{3}i.
$$

The complete factorization of $f(x)$ is

$$
(x - 1)^3 + 1 = x^3 - 9x^2 + 27x - 26 = (x - 2)\left(x - \frac{7}{2} - \frac{1}{2}\sqrt{3}i\right)\left(x - \frac{7}{2} + \frac{1}{2}\sqrt{3}i\right).
$$

42. For $x + \frac{1}{2}$ to be a factor of $f(x) = 8x^2 - 4kx + 9$ we need

$$f\left(-\frac{1}{2}\right) = 8\left(-\frac{1}{2}\right)^2 - 4k\left(-\frac{1}{2}\right) + 9 = 2 + 2k + 9 = 11 + 2k = 0$$

or $k = -\frac{11}{2}$.

45. The partial fraction decomposition looks like

$$\frac{2x - 1}{x(x^2 + 2x - 3)} = \frac{2x - 1}{x(x + 3)(x - 1)} = \frac{A}{x} + \frac{B}{x + 3} + \frac{C}{x - 1}.$$

Multiplying both sides of this equality by the denominator $x(x+3)(x-1)$, we obtain

$$2x - 1 = A(x + 3)(x - 1) + Bx(x - 1) + Cx(x + 3). \qquad (3.21)$$

Since the denominator consists solely of linear factors, none of which is repeated, we can use the *Shortcut Worth Knowing* described on page 174 in Section 3.6 of the text. Letting $x = 0$ in (3.21) we have

$$2(0) - 1 = -1 = A(3)(-1) + B(0) + C(0) = -3A,$$

so $A = \frac{1}{3}$. Letting $x = -3$ in (3.21) we have

$$2(-3) - 1 = -7 = A(0) + B(-3)(-3 - 1) + C(0) = 12B,$$

so $B = -\frac{7}{12}$. Letting $x = 1$ in (3.21) we have

$$2(1) - 1 = 1 = A(0) + B(0) + C(1)(1 + 3) = 4C,$$

so $C = \frac{1}{4}$. The partial fraction decomposition is

$$\frac{2x - 1}{x(x^2 + 2x - 3)} = \frac{1/3}{x} + \frac{-7/12}{x + 3} + \frac{1/4}{x - 1}.$$

48. Since the degree of the numerator is greater than the degree of the denominator, we first use polynomial long division:

$$
\begin{array}{r}
x^3 + x^2 + 3x + 5 \\
x^2 - 2x + 1 \overline{\smash{)}\ x^5 - x^4 + 2x^3 + 0x^2 + 5x - 1} \\
\underline{x^5 - 2x^4 + x^3} \\
x^4 + x^3 + 0x^2 + 5x - 1 \\
\underline{x^4 - 2x^3 + x^2} \\
3x^3 - x^2 + 5x - 1 \\
\underline{3x^3 - 6x^2 + 3x} \\
5x^2 + 2x - 1 \\
\underline{5x^2 - 10x + 5} \\
12x - 6
\end{array}
$$

Thus,

$$\frac{x^5 - x^4 + 2x^3 + 5x - 1}{(x-1)^2} = x^3 + x^2 + 3x + 5 + \frac{12x - 6}{(x-1)^2}.$$

The partial fraction decomposition of $(12x - 6)/(x-1)^2$ looks like

$$\frac{12x - 6}{(x-1)^2} = \frac{A}{x-1} + \frac{B}{(x-1)^2}.$$

Multiplying both sides of this equality by the denominator $(x-1)^2$, we obtain

$$12x - 6 = A(x - 1) + B \qquad (3.22)$$
$$\text{or} \qquad 12x - 6 = Ax + (-A + B). \qquad (3.23)$$

Letting $x = 1$ in (3.22) we find

$$12(1) - 6 = 6 = A(0) + B,$$

so $B = 6$. Equation (3.23) then becomes

$$12x - 6 = Ax + (-A + 6).$$

Equating the coefficients of x, we see that $A = 12$. Thus, the partial fraction decomposition of the original rational expression is

$$\frac{x^5 - x^4 + 2x^3 + 5x - 1}{(x-1)^2} = x^3 + x^2 + 3x + 5 + \frac{12}{x-1} + \frac{6}{(x-1)^2}.$$

51. Since

$$f(x) = \frac{2x}{x^2 + 1}$$

has no vertical asymptotes and its horizontal asymptote is $y = 0$, its graph must be either (a) or (f). Because $f(x)$ is negative for $x < 0$, it must be (f).

54. Since

$$f(x) = 2 - \frac{1}{x^2} = \frac{2x^2 - 1}{x^2}$$

has vertical asymptote $x = 0$ (the y-axis), its graph must be (g).

57. Since

$$f(x) = \frac{x^2 - 10}{2x - 4} = \frac{1}{2} \cdot \frac{x^2 - 10}{x - 2} = \frac{1}{2}\left(x + 2 - \frac{6}{x - 2}\right)$$

(using synthetic division), we see that the graph of $f(x)$ has vertical asymptote $x = 2$ and slant asymptote $y = \frac{1}{2}x + 1$. Thus, the graph of f must be (c).

60. Since

$$f(x) = \frac{3}{x^2 + 1}$$

has no vertical asymptotes and its horizontal asymptote is $y = 0$, the graph must be either (a) or (f). Because $f(x)$ is always positive, it must be (a).

Chapter 4

Trigonometric Functions

4.1 | Angles and Their Measurement |

3.

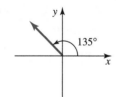

6.

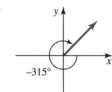

9.

12.

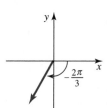

15.

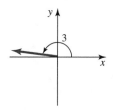

18. To convert from degrees to radians we use

$$1° = 15(1°) = 15 \cdot \left(\frac{\pi}{180} \text{ radian} \right) = \frac{\pi}{12} \text{ radian.}$$

21. To convert from degrees to radians we use

$$270° = 270(1°) = 270 \cdot \left(\frac{\pi}{180} \text{ radian} \right) = \frac{3\pi}{2} \text{ radians.}$$

24. To convert from degrees to radians we use

$$540° = 540(1°) = 540 \cdot \left(\frac{\pi}{180} \text{ radian} \right) = 3\pi \text{ radians.}$$

27. To convert from radians to degrees we use

$$\frac{2\pi}{3} \text{ radians} = \frac{2\pi}{3}(1 \text{ radian}) = \frac{2\pi}{3} \left(\frac{180}{\pi} \right)^° = 120°.$$

30. To convert from radians to degrees we use

$$7\pi \text{ radians} = 7\pi(1 \text{ radian}) = 7\pi \left(\frac{180}{\pi} \right)^° = 1260°.$$

33. (a) Since $875 = 2(360) + 155$, we see that $875°$ is coterminal with $155°$.

    (b) Subtracting $155°$ from $360°$, we see that the angle is coterminal with $-205°$.

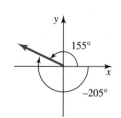

36. (a) Since $-610 - 2(360) + 110$, we see that $-610°$ is coterminal with $110°$.

(b) Subtracting $110°$ from $360°$, we see that the angle is coterminal with $250°$.

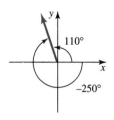

39. (a) In this case we want to know how many integer multiples of 2π there are in 5.3π. The answer is easily seen to be 2 with a remainder of $5.3\pi - 2(2\pi) = 1.3\pi$. That is, $5.3\pi = 2(2\pi) + 1.3\pi$, so 5.3π is coterminal with 1.3π.

(b) Subtracting 1.3π from 2π, we see that the angle is coterminal with -0.7π.

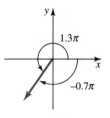

42. (a) In this case, we want to know how many integer multiples of 2π there are in 7.5. The answer, obtained by dividing 7.5 by 2π is 1 with a remainder of 1.21681. That is, $7.5 = 1(2\pi) + 1.21681$.

(b) Subtracting 1.21681 from 2π, we see that the angle is coterminal with 5.06638.

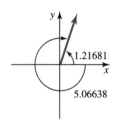

45. (a) Since $98.4° > 90°$, it is not an acute angle and hence, has no complementary angle.

(b) Since the two positive angles are supplementary if their sum is $180°$, the angle supplementary to $98.4°$ is $180° - 98.4° = 81.6°$.

48. (a) Since two acute angles are complementary if their sum is $\pi/2$ radians, the angle complementary to $\pi/6$ is

$$\frac{\pi}{2} - \frac{\pi}{6} = \frac{3\pi}{6} - \frac{\pi}{6} = \frac{2\pi}{6} = \frac{\pi}{3} \text{ radians.}$$

(b) Since two positive angles are supplementary if their sum is π radians, the angle supplementary to $\pi/6$ is

$$\pi - \frac{\pi}{6} = \frac{6\pi}{6} - \frac{\pi}{6} = \frac{5\pi}{6} \text{ radians.}$$

51. **(a)** A counter-clockwise rotation is $360°$, so three-fifths of a clockwise rotation is

$$\frac{3}{5}(360°) = 216° \quad \text{or} \quad 216(1°) = 216\left(\frac{\pi}{180} \text{ radians}\right) = \frac{6\pi}{5} \text{ radians}.$$

   (b) Since $5\frac{1}{8} = \frac{41}{8}$, five and one-eighth clockwise rotations is

$$\frac{41}{8}(-360°) = -1845°$$

   or

$$(-1845)(1°) = (-1845)\left(\frac{\pi}{180} \text{ radians}\right) = -\frac{41\pi}{4} \text{ radians}.$$

54. In 2 hours, the minute hand on a clock makes 2 clockwise rotations. This is $2(-360°) = -720°$ or $2(-2\pi) = -4\pi$ radians.

57. The arc length subtended by an angle of 3 radians is $s = r(3) = 3r$, where r is the radius of the circle.

   (a) When $r = 3$, $s = 3(3) = 9$.

   (b) When $r = 5$, $s = 3(5) = 15$.

60. Using $s = r\theta$ where θ is measured in radians, we have $\pi/3 = \theta$ when the radius is 1 and the central angle is $\pi/3$. Thus, **(a)** $\theta = \pi/3$ radian or **(b)** $(\pi/3)(1 \text{ radian}) = (\pi/3)(180/\pi)° = 60°$.

63. We use the fact that $1' = \left(\frac{1}{60}\right)°$ and $1'' = \frac{1}{60}\left(\frac{1}{60}\right)° = \left(\frac{1}{3600}\right)°$.

   (a) $5°10' = 5° + 10(1') = 5° + 10\left(\frac{1}{60}\right)° = \left(5 + \frac{10}{60}\right)° = 5.166667°$

   (b) $10°25' = 10° + 25(1') = 10° + 25\left(\frac{1}{60}\right)° = \left(10 + \frac{25}{60}\right)° = 10.416667°$

   (c) $10°39'17'' = 10° + 39(1') + 17(1'') = 10° + 39\left(\frac{1}{60}\right)° + 17\left(\frac{1}{3600}\right)° = \left(10 + \frac{39}{60} + \frac{17}{3600}\right)°$
   $= 10.654722°$

   (d) $143°7'2'' = 143° + 7(1') + 2(1'') = 143° + 7\left(\frac{1}{60}\right)° + 2\left(\frac{1}{3600}\right)° = \left(143 + \frac{7}{60} + \frac{2}{3600}\right)°$
   $= 143.11722°$

66. **(a)** We first convert $7.2°$ to radians:

$$7.2° = 7.2(1°) = 7.2\left(\frac{\pi}{180} \text{ radians}\right) = 0.125664 \text{ radian}.$$

   Since arc length is given by $s = r\theta$, we have

$$5000 = r(0.125664) \quad \text{or} \quad r = 39788.7 \text{ stades}.$$

(b) Since 1 stade = 559 feet, we have that Eratosthenes' measure of the radius of the earth is

$$39788.7(1 \text{ stade}) = 39788.7(559 \text{ feet}) = 22{,}241{,}900 \text{ feet}.$$

The circumference of the earth would then be

$$2\pi(22{,}241{,}900 \text{ feet}) = 139{,}750{,}000 \text{ feet} = 139{,}750{,}000(1 \text{ foot})$$

$$= 139{,}750{,}000 \left(\frac{1}{5280} \text{ mi} \right) = 26{,}467.8 \text{ mi}.$$

If the actual diameter of the earth is 7900 miles, then its circumference is $7900\pi = 24{,}818.6$ miles. Thus, Eratosthenes' value is in error by $26{,}467.8 - 24{,}818.6 = 1649.22$ miles. The error as a percentage of the actual circumference is

$$\frac{1649.22}{24{,}818.6} \times 100\% = 6.6\%.$$

69. **(a)** Since the knot makes 6 revolutions in 4 seconds, it travels through $6(2\pi) = 12\pi$ radians, so its angular speed is $12\pi/4 = 3\pi$ radians per second.

(b) The yo-yo is on the arc and the other end of the string is at the center, so the radius of the arc is $100 - 40 = 60$ cm. In 1 revolution it travels $60(2\pi) = 120\pi$ cm, and in 6 revolutions it travels $6(120\pi) = 720\pi$ cm. Its linear speed is then

$$\text{speed} = \frac{\text{distance}}{\text{time}} = \frac{720\pi \text{ cm}}{4 \text{ s}} = 180\pi \text{ cm/s}.$$

4.2   The Sine and Cosine Functions

3. Using the Pythagorean Identity gives $(-2/3)^2 + \cos^2 t = 1$ or $\cos^2 t = 5/9$. Since $P(t)$ is in the third quadrant, $\cos t < 0$, so $\cos t = -\sqrt{5}/3$.

6. Using the Pythagorean Identity gives $\sin^2 t + (3/10)^2 = 1$ or $\sin^2 t = 91/100$. Thus, $\sin t = \pm\sqrt{91}/10$.

9. Since $2\sin t - \cos t = 0$, $\cos t = 2\sin t$, and from the Pythagorean Identity we have

$$\sin^2 t + \cos^2 t = 1$$
$$\sin^2 t + (2\sin t)^2 = 1$$
$$5\sin^2 t = 1$$
$$\sin^2 t = \frac{1}{5}.$$

Thus, $\sin t = \pm 1/\sqrt{5}$ and $\cos t = 2\sin t = \pm 2/\sqrt{5}$.

12. **(a)** Since $\sin 3\pi = \sin(\pi + 2\pi) = \sin \pi$, we have from Figure 4.2.4 in the text that $\sin 3\pi = 0$.

(b) Similarly, $\cos 3\pi = \cos \pi = -1$.

15. The angle $t = 2\pi/3$ is in the second quadrant and the reference angle is $t' = \pi - 2\pi/3 = \pi/3$, so

$$\sin\frac{2\pi}{3} = \sin\frac{\pi}{3} = \frac{\sqrt{3}}{2} \quad \text{and} \quad \cos\frac{2\pi}{3} = -\cos\frac{\pi}{3} = -\frac{1}{2}.$$

18. The angle $t = 3\pi/4$ is in the second quadrant and the reference angle is $t' = \pi - 3\pi/4 = \pi/4$, so

$$\sin\frac{3\pi}{4} = \sin\frac{\pi}{4} = \frac{\sqrt{2}}{2} \quad \text{and} \quad \cos\frac{3\pi}{4} = -\cos\frac{\pi}{4} = -\frac{\sqrt{2}}{2}.$$

21. The angle $t = -\pi/4$ is in the fourth quadrant and the reference angle is $t' = -(-\pi/4) = \pi/4$, so

$$\sin\left(-\frac{\pi}{4}\right) = -\sin\frac{\pi}{4} = -\frac{\sqrt{2}}{2} \quad \text{and} \quad \cos\left(-\frac{\pi}{4}\right) = \cos\frac{\pi}{4} = \frac{\sqrt{2}}{2}.$$

24. The angle $t = -11\pi/6$ is in the first quadrant and the reference angle is $t' = 2\pi - 11\pi/6 = \pi/6$, so

$$\sin\left(-\frac{11\pi}{6}\right) = \sin\frac{\pi}{6} = \frac{1}{2} \quad \text{and} \quad \cos\left(-\frac{11\pi}{6}\right) = \cos\frac{\pi}{6} = \frac{\sqrt{3}}{2}.$$

27. The angle $t = -11\pi/3$ is in the first quadrant and the reference angle is $t' = 4\pi - 11\pi/3 = \pi/3$, so

$$\sin\left(-\frac{11\pi}{3}\right) = \sin\frac{\pi}{3} = \frac{\sqrt{3}}{2}.$$

30. Since $-19\pi/2 = -20\pi/2 + \pi/2$,

$$\sin\left(-\frac{19\pi}{2}\right) = \sin\left(-\frac{20\pi}{2} + \frac{\pi}{2}\right) = \sin\left(-10\pi + \frac{\pi}{2}\right) = \sin\frac{\pi}{2} = 1.$$

33. Using periodicity,

$$\sin 3\pi = \sin(\pi + 2\pi) = \sin\pi.$$

36. Using periodicity,

$$\cos 16.8\pi = \cos(14.8\pi + 2\pi) = \cos 14.8\pi.$$

39. The angle $t = 135°$ is in the second quadrant and the reference angle is $t = 180° - 135° = 45°$, so

$$\sin 135° = \sin 45° = \frac{\sqrt{2}}{2}.$$

42. The terminal side of $t = 270°$ lies along the negative y-axis, so the point $P(t)$ is $(0, -1)$ and $\sin 270° = -1$.

45. The angle $t = -60°$ is in the fourth quadrant and the reference angle is $60°$, so

$$\sin(-60°) = -\sin 60° = -\frac{\sqrt{3}}{2}.$$

48. Since $\cos t$ is the x-coordinate of $P(t)$, $\cos t = 1$ only for $t = 3\pi/2$ in $[0, 2\pi)$.

51. Since $\cos 30° = \sqrt{3}/2$, the only other angle in $[0°, 360°)$ whose cosine is $\sqrt{3}/2$ lies in the fourth quadrant and has reference angle $30°$. This angle is $330°$.

54. Since $\cos\theta$ is the x-coordinate of $P(\theta)$, $\cos\theta = 1$ only for $\theta = 90°$ in $[0°, 360°)$.

57. (a) When $\theta = 0$, $\sin\theta = 0$ and $\sin 2\theta = 0$, so

$$g_{\text{sat}} = 978.0309 \text{ cm/}s^2.$$

    (b) At the north pole, $\theta = 90°$, and $\sin\theta = 1$ and $\sin 2\theta = \sin 180° = 0$, so

$$g_{\text{sat}} = 978.0309 + 5.18552 = 983.2164 \text{ cm/}s^2.$$

    (c) When $\theta = 45°$, $\sin\theta = \sqrt{2}/2$ and $\sin 2\theta = \sin 90° = 1$, so

$$g_{\text{sat}} = 978.0309 + 5.18552(\sqrt{2}/2)^2 - 0.00570(1)^2$$
$$= 978.0309 + 2.59276 - 0.00570 = 980.618 \text{ cm/}s^2.$$

4.3   Graphs of Sine and Cosine Functions

3. The period (and hence the length of one cycle) of $y = 2 - \sin x$ is 2π. The graph of this function, shown in blue, can be obtained from the graph of $y = \sin x$, shown in black, by reflection through the x-axis and a vertical upward shift of 2.

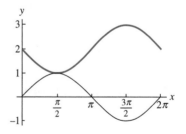

6. The period (and hence the length of one cycle) of $y = 1 - 2\sin x$ is $2\pi/2 = \pi$. The graph of this function, shown in blue, can be obtained from the graph of $y = \sin x$, shown in black, by reflection through the x-axis, vertical stretching by a factor of 2, and a vertical upward shift of 1.

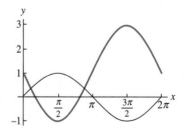

9. Since the graph flattens out at $x = 0$, we use $y = A\cos x + D$. The amplitude is $\frac{1}{2}[4 - (-2)] = 3$ and the graph has been reflected through the line $y = 1$, so $A = -3$ and $D = 1$. Thus, $y = -3\cos x + 1$.

12. Setting $-\cos 2x = 0$, we have by (2) in this section of the text that the x-intercepts are determined by $2x = (2n+1)\pi/2$, n an integer. Thus, the x-intercepts are at $x = (2n+1)\pi/4$, so the intercepts are $(\pi/4 + n\pi/2, 0)$, where n is an integer.

15. Setting $\sin(x - \pi/4) = 0$ we have by (1) in this section of the text that the x-intercepts are determined by $x - \pi/4 = n\pi$, n an integer. Thus, the x-intercepts are at $x = n\pi + \pi/4 = (n + \frac{1}{4})\pi$, so the intercepts are $(\pi/4 + n\pi, 0)$, where n is an integer.

18. Setting $1 - 2\cos x = 0$ we have $2\cos x = 1$ or $\cos x = \frac{1}{2}$. Since $\cos x$ is positive, x is in the first or fourth quadrant. Since $\cos x = \frac{1}{2}$ for $x = \pi/3$, we see that $1 - 2\cos x = 0$ for $x = \pi/3$ and $x = 5\pi/3$ in $[0, 2\pi]$. By periodicity, then, the x-intercepts of $1 - 2\cos 2x$ are at $\pi/3 + 2n\pi = (\frac{1}{3} + 2n)\pi$ and $5\pi/3 + 2n\pi = (\frac{5}{3} + 2n)\pi$, so the intercepts are $(\pi/3 + 2n\pi, 0)$ and $(5\pi/3 + 2n\pi, 0)$, where n is an integer.

21. Since the y-intercept is $\frac{1}{2}$ (not 0), the equation has the form $y = A\cos Bx$. The amplitude of the graph is $A = \frac{1}{2}$ and the period is $2 = 2\pi/B$, so $B = \pi$ and $y = \frac{1}{2}\cos \pi x$.

24. Since the y-intercept is 3 (not 0), the equation has the form $y = A\cos Bx$. The amplitude of the graph is $A = 3$ and the period is $8 = 2\pi/B$, so $B = \pi/4$ and $y = 3\cos \pi x/4$.

27. The amplitude of $y = -3\cos 2\pi x$ is $A = |-3| = 3$ and the period is $2\pi/2\pi = 1$.

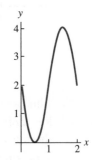

30. The amplitude of $y = 2 - 2\sin \pi x$ is $A = |-2| = 2$ and the period is $2\pi/\pi = 2$.

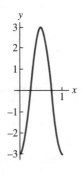

33. The amplitude of $y = \sin(x - \pi/6)$ is $A = 1$ and the period is $2\pi/1 = 2\pi$. The phase shift is $|-\pi/6|/1 = \pi/6$. Since $C = -\pi/6 < 0$, the shift is to the right.

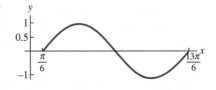

36. The amplitude of $y = -2\cos(2x - \pi/6)$ is $A = |-2| = 2$ and the period is $2\pi/2 = \pi$. The phase shift is $|-\pi/6|/2 = -\pi/12$. Since $C = -\pi/6 < 0$, the shift is to the right.

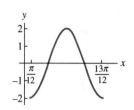

39. The amplitude of $y = 3\sin(x/2 - \pi/3)$ is $A = 3$ and the period is $2\pi/(1/2) = 4\pi$. The phase shift is $|-\pi/3|/(1/2) = 2\pi/3$. Since $C = -\pi/3 < 0$, the shift is to the right.

42. The amplitude of $y = 2\cos(-2\pi x - 4\pi/3) = 2\cos(2\pi x + 4\pi/3)$ is $A = 2$ and the period is $2\pi/2\pi = 1$. The phase shift is $|4\pi/3|/2\pi = \frac{2}{3}$. Since $C = 4\pi/3 > 0$, the shift is to the left.

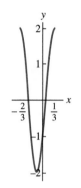

45. We identify $A = 3$ and $2\pi/B = 2\pi/3$, so $B = 3$. Since the graph is shifted to the right, $C < 0$. The phase shift is $|C|/B = |C|/3 = \pi/3$, so $C = -\pi$. The functions are $y = 3\sin(3x - \pi)$, shown as the dashed curve, and $y = 3\cos(3x - \pi)$, shown as the solid blue curve.

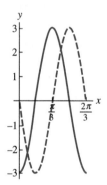

48. We identify $A = \frac{5}{4}$ and $2\pi/B = 4$, so $B = \pi/2$. Since the graph is shifted to the left, $C > 0$. The phase shift is $|C/B| = C/(\pi/2) = 1/2\pi$, so $C = \frac{1}{4}$. The functions are $y = \frac{5}{4}\sin(\pi x/2 + 1/2\pi)$, shown as the dashed curve, and $y = \frac{5}{4}\cos(\pi x/2 + 1/2\pi)$, shown as the solid blue curve.

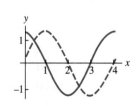

51. For $\omega = 2$ and $\theta_0 = \pi/10$ the function is $\theta(t) = (\pi/10)\cos 2t$. The amplitude is $A = 2$ and the period is $2\pi/2 = \pi$. Thus, two cycles will extend from $t = 0$ to $t = 2\pi$.

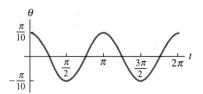

54. **(a)** Since 8 A.M. is 8 hours after midnight,

$$T(8) = 50 + 10\sin \pi(8 - 8)/12 = 50$$

and the temperature at 8 A.M. is $50°$.

(b) We solve

$$50 + 10\sin \frac{\pi}{12}(t - 8) = 60$$

$$\sin \frac{\pi}{12}(t - 8) = 1$$

$$\frac{\pi}{12}(t - 8) = \frac{\pi}{2}$$

$$t - 8 = 6$$

$$t = 14.$$

Since 14 hours after midnight is noon plus 2 hours, the temperature will be $60°$F at 2 P.M.

(c) The amplitude of the graph is $A = 10$ and the graph is vertically centered at $T = 50$. The period is $2\pi/(\pi/12) = 24$ and the phase shift is $|C|/B = (8\pi/12)/(\pi/12) = 8$. Since $C = 8\pi/12 > 0$, the graph of $y = 50 + 10\sin \pi t/12$ is shifted 8 units to the left to obtain the graph of

$$T = 50 + 10\sin(\pi t/12 - 8\pi/12).$$

(d) The minimum temperature will be $40°$F and this will occur when $\pi(t - 8)/12 = 3\pi/2$, since $x = 3\pi/2$ where $y = \sin x$ first reaches its minimum. Solving for t, we obtain $t = 26$. Since the period is 24, the temperature is first minimum when $t = 2$ or at 2 A.M. The maximum temperature will be $60°$F and this will occur when $\pi(t - 8)/12 = \pi/2$, since $x = \pi/2$ where $y = \sin x$ first reaches its maximum. Solving for t, we obtain $t = 14$. Thus, the temperature is first maximum when $t = 14$ or at 2 P.M. (Note that we could have reached the same conclusion about the first occurrence of the maximum temperature from the information in parts (b) and (c).)

4.4 $\boxed{\textbf{Other Trigonometric Functions}}$

3. $\cot \dfrac{13\pi}{6} = \cot\left(\dfrac{13\pi}{6} - 2\pi\right)$ (because the cotangent is π-periodic)

$\qquad = \cot \dfrac{\pi}{6} = \dfrac{\cos \frac{\pi}{6}}{\sin \frac{\pi}{6}}$ (by the definition of the cotangent)

$\qquad = \dfrac{\sqrt{3}/2}{1/2} = \sqrt{3}$

6. $\sec 7\pi = \sec(7\pi - 6\pi)$ (because the secant is 2π-periodic)

$\qquad = \sec \pi = \dfrac{1}{\cos \pi}$ (by the definition of the secant)

$\qquad = \dfrac{1}{-1} = -1$

9. $\tan \dfrac{23\pi}{4} = \tan\left(\dfrac{23\pi}{4} - 5\pi\right)$ (because the tangent is π-periodic)

$\qquad = \tan \dfrac{3\pi}{4} = \dfrac{\sin \frac{3\pi}{4}}{\cos \frac{3\pi}{4}}$ (by the definition of the tangent)

$\qquad = \dfrac{\sqrt{2}/2}{-\sqrt{2}/2} = -1$

12. $\cot \dfrac{17\pi}{6} = \cot\left(\dfrac{17\pi}{6} - 2\pi\right)$ (because the cotangent is π-periodic)

$\qquad = \cot \dfrac{5\pi}{6} = \dfrac{\cos \frac{5\pi}{6}}{\sin \frac{5\pi}{6}}$ (by the definition of the cotangent)

$\qquad = \dfrac{-\sqrt{3}/2}{1/2} = -\sqrt{3}$

15. $\sec(-120°) = \sec(-120° + 360°)$ (because the secant has period $360°$)

$\qquad = \sec 240° = \dfrac{1}{\cos 240°}$ (by the definition of the secant)

$\qquad = \dfrac{1}{-1/2} = -2$

18. $\cot(-720°) = \cot(-720° + 2 \cdot 360°)$ (because the cotangent has period $360°$)

$\qquad = \cot 0° = \dfrac{\cos 0°}{\sin 0°}$ (by the definition of the cotangent)

Since $\sin 0°$ is 0, the cotangent of $-720°$ is not defined.

21. We first compute $\cot x$ using the Pythagorean identity:

$$\cot^2 x = \csc^2 x - 1 = \left(\dfrac{4}{3}\right)^2 + 1 = \dfrac{16}{9} - 1 = \dfrac{7}{9}.$$

Since $0 < x < \pi/2$, x is in the first quadrant and $\cot x > 0$. Thus, $\cot x = \sqrt{7}/3$.

Then

$$\tan x = \frac{1}{\cot x} = \frac{1}{\sqrt{7}/3} = \frac{3}{\sqrt{7}}.$$

Finally,

$$\sin x = \frac{1}{\csc x} = \frac{1}{4/3} = \frac{3}{4},$$

$$\cos x = \frac{\cos x \sin x}{\sin x} = \cot x \sin x = \frac{\sqrt{7}}{3}\left(\frac{3}{4}\right) = \frac{\sqrt{7}}{4},$$

and

$$\sec x = \frac{1}{\cos x} = \frac{1}{\sqrt{7}/4} = \frac{4}{\sqrt{7}}.$$

24. We first compute $\sin x$ using the Pythagorean identity:

$$\sin^2 x = 1 - \cos^2 x = 1 - \left(-\frac{1}{\sqrt{5}}\right)^2 = 1 - \frac{1}{5} = \frac{4}{5}.$$

Since $\pi < x < 3\pi/2$, x is in the third quadrant and $\sin x < 0$. Thus, $\sin x = -2/\sqrt{5}$. Then

$$\tan x = \frac{\sin x}{\cos x} = \frac{-2/\sqrt{5}}{-1/\sqrt{5}} = 2.$$

Finally,

$$\cot x = \frac{1}{\tan x} = \frac{1}{2},$$

$$\sec x = \frac{1}{\cos x} = \frac{1}{-1/\sqrt{5}} = -\sqrt{5},$$

and

$$\csc x = \frac{1}{\sin x} = \frac{1}{-2/\sqrt{5}} = -\frac{\sqrt{5}}{2}.$$

27. From $3\cos x = \sin x$ we see that $\tan x = \sin x/\cos x = 3$. Since $\tan x = 3 > 0$, x is in either the first or third quadrant. (This is consistent with the fact that $\sin x$ and $\cos x$ have the same algebraic sign, as can be inferred from $3\cos x = \sin x$.) From $\tan x = 3$ we see that $\cot x = 1/\tan x = 1/3$. Using the Pythagorean identity, we have

$$\sec^2 x = 1 + \tan^2 x = 1 + 3^2 = 10,$$

so $\sec x = \pm\sqrt{10}$. From

$$\csc^2 x = 1 + \cot^2 = 1 + \left(\frac{1}{3}\right)^2 = \frac{10}{9}$$

we see that $\csc x = \pm\sqrt{10}/3$. Finally, $\sin x = 1/\csc x = 1/(\pm\sqrt{10}/3) = \pm 3/\sqrt{10}$, and $\cos x = 1/\sec x = \pm 1/\sqrt{10}$.

30. The period of $y = \tan(x/2)$ is $\pi/(1/2) = 2\pi$. Since

$$\tan(x/2) = \sin(x/2)/\cos(x/2),$$

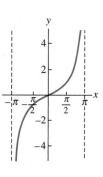

the x-intercepts of $\tan(x/2)$ occur at the zeros of $\sin(x/2)$; namely, at $x/2 = n\pi$ or $x = 2n\pi$ for n an integer. The vertical asymptotes occur at the zeros of $\cos(x/2)$; namely, at $x/2 = (2n+1)\pi/2$ or $x = (2n+1)\pi = \pi + 2n\pi$ for n an integer. Since the graph has vertical asymptotes at $-\pi$ and π (using $n = -1$ and $n = 0$), we graph one cycle on the interval $(-\pi, \pi)$.

33. The period of $y = \tan(x/2 - \pi/4)$ is $\pi/(1/2) = 2\pi$. Since

$$\tan(x/2 - \pi/4) = \sin(x/2 - \pi/4)/\cos(x/2 - \pi/4),$$

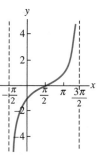

the x-intercepts of $\tan(x/2 - \pi/4)$ occur at the zeros of $\sin(x/2 - \pi/4)$; namely, at $x/2 - \pi/4 = n\pi$ or $x = 2n\pi + \pi/2$ for n an integer. The vertical asymptotes occur at the zeros of $\cos(x/2 - \pi/4)$; namely, at $x/2 - \pi/4 = (2n+1)\pi/2$ or $x = (2n+1)\pi + \pi/2 = 3\pi/2 + 2n\pi$ for n an integer. Since the graph has vertical asymptotes at $-\pi/2$ and $3\pi/2$ (using $n = -1$ and $n = 0$), we graph one cycle on the interval $(-\pi/2, 3\pi/2)$.

36. The period of $y = \tan(x + 5\pi/6)$ is π. Since $\tan(x + 5\pi/6) = \sin(x + 5\pi/6)/\cos(x + 5\pi/6)$, the x-intercepts of $\tan(x + 5\pi/6)$ occur at the zeros of $\sin(x + 5\pi/6)$; namely, at $x + 5\pi/6 = n\pi$ or $x = n\pi - 5\pi/6$ for n an integer. The vertical asymptotes occur at the zeros of $\cos(x + 5\pi/6)$; namely, at $x + 5\pi/6 = (2n+1)\pi/2$ or $x = (2n+1)\pi/2 - 5\pi/6 = -\pi/3 + n\pi$, for n an integer. Since the graph has vertical asymptotes at $-\pi/3$ and $2\pi/3$ (using $n = 0$ and $n = 1$), we graph one cycle on the interval $(-\pi/3, 2\pi/3)$.

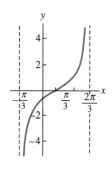

39. The period of $y = 3\csc\pi x$ is $2\pi/\pi = 2$. Since $3\csc\pi x = 3/\sin\pi x$, the vertical asymptotes occur at the zeros of $\sin\pi x$; namely, at $\pi x = n\pi$ or $x = n$, for n an integer. We plot one cycle of the graph on $(-1, 1)$, since the period of the function is $2 = 1 - (-1)$ and vertical asymptotes occur at $x = -1$ and $x = 1$ (taking $n = -1$ and 1).

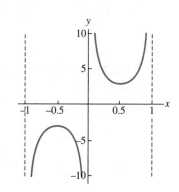

42. The period of $y = \csc(4x+\pi)$ is $2\pi/4 = \pi/2$. Since $\csc(4x + \pi) = 1/\sin(4x + \pi)$, the vertical asymptotes occur at the zeros of $\sin(4x + \pi)$, namely, at $4x + \pi = n\pi$ or $x = (n-1)\pi/4 = -\pi/4 + n\pi/4$, for n an integer. We plot one cycle of the graph on $(-\pi/4, \pi/4)$ since the period of the function is $\pi/2 = \pi/4 - (-\pi/4)$ and vertical asymptotes occur at $x = -\pi/4$ and $x = \pi/4$ (taking $n = 0$ and 2).

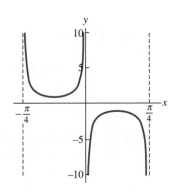

45. The graph of $y = \cot x$ is shown in the figures below as a dashed curve, while the graphs of $y = A\tan(x + C)$, for various choices of A and C, are shown as solid blue curves.

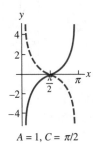 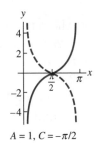 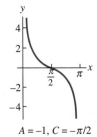

$$A = 1, \; C = \pi/2 \qquad A = 1, \; C = -\pi/2 \qquad A = -1, \; C = -\pi/2$$

We see from the third graph that $\cot x = -\tan(x - \pi/2)$.

4.5   Special Identities

3. Here, we use the Pythagorean identity $1 + \tan^2\theta = \sec^2\theta$. With $x = a\sec\theta$, and assuming $a > 0$, we have

$$\sqrt{x^2 - a^2} = \sqrt{(a\sec\theta)^2 - a^2} = \sqrt{a^2\sec^2\theta - a^2} = a\sqrt{\sec^2\theta - 1}$$
$$= a\sqrt{1 + \tan^2\theta - 1} = a\sqrt{\tan^2\theta} = a|\tan\theta|.$$

Since $\tan\theta \geq 0$ for $0 \leq \theta < \pi/2$, $|\tan\theta| = \tan\theta$ and

$$\sqrt{x^2 - a^2} = a\tan\theta.$$

6. Here, we use the Pythagorean identity $1 + \tan^2\theta = \sec^2\theta$. With $x = \sqrt{3}\sec\theta$ we have $x^2 = 3\sec^2\theta$ and

$$\frac{\sqrt{x^2 - 3}}{x^2} = \frac{\sqrt{(\sqrt{3}\sec\theta)^2 - 3}}{3\sec^2\theta} = \frac{\sqrt{3\sec^2\theta - 3}}{3\sec^2\theta} = \frac{\sqrt{3 + 3\tan^2\theta - 3}}{3\sec^2\theta}$$
$$= \frac{\sqrt{3}\tan\theta}{3\sec^2\theta} = \frac{\sqrt{3}\sin\theta/\cos\theta}{3/\cos^2\theta} = \frac{1}{3}\sin\theta\cos\theta.$$

In the previous equations we used $|\tan^2\theta| = \tan\theta$ because $0 < \theta < \pi/2$.

9. Since $\pi/12 = \pi/3 - \pi/4$,

$$\cos\frac{\pi}{12} = \cos\left(\frac{\pi}{3} - \frac{\pi}{4}\right) = \cos\frac{\pi}{3}\cos\frac{\pi}{4} + \sin\frac{\pi}{3}\sin\frac{\pi}{4}$$

$$= \frac{1}{2}\frac{\sqrt{2}}{2} + \frac{\sqrt{3}}{2}\frac{\sqrt{2}}{2} = \frac{\sqrt{2}}{4}(1 + \sqrt{3}).$$

Since $\sqrt{2}\sqrt{3} = \sqrt{6}$, we also have $\cos(\pi/12) = (\sqrt{2} + \sqrt{6})/4$.

12. Since $75° = 30° + 45°$,

$$\cos 75° = \cos(30° + 45°) = \cos 30° \cos 45° - \sin 30° \sin 45°$$

$$= \frac{\sqrt{3}}{2}\frac{\sqrt{2}}{2} - \frac{1}{2}\frac{\sqrt{2}}{2} = \frac{\sqrt{6} - \sqrt{2}}{4}.$$

Since $\sqrt{6} = \sqrt{3}\sqrt{2}$ we also have $\cos 75° = (\sqrt{3} - 1)/2\sqrt{2}$.

15. Since $5\pi/12 = \pi/6 + \pi/4$ (or $5\pi/12 = 75° = 30° + 45° = \pi/6 + \pi/4$),

$$\tan\frac{5\pi}{12} = \frac{\tan(\pi/6) + \tan(\pi/4)}{1 - \tan(\pi/6)\tan(\pi/4)} = \frac{1/\sqrt{3} + 1}{1 - (1/\sqrt{3})(1)} = \frac{1/\sqrt{3} + 1}{1 - 1/\sqrt{3}}\left(\frac{\sqrt{3}}{\sqrt{3}}\right)$$

$$= \frac{1 + \sqrt{3}}{\sqrt{3} - 1} = 2 + \sqrt{3}.$$

18. Since $11\pi/12 = 165° = 135° + 30°$,

$$\tan\frac{11\pi}{12} = \tan 165° = \frac{\tan 135° + \tan 30°}{1 - \tan 135° \tan 30°} = \frac{-1 + 1/\sqrt{3}}{1 - (-1)(1/\sqrt{3})}$$

$$= \frac{-1 + 1/\sqrt{3}}{1 + 1/\sqrt{3}}\left(\frac{\sqrt{3}}{\sqrt{3}}\right) = \frac{1 - \sqrt{3}}{1 + \sqrt{3}}.$$

21. Since $165° = 135° + 30°$,

$$165° = \cos 135° \cos 30° - \sin 135° \sin 30° = \left(-\frac{\sqrt{2}}{2}\right)\left(\frac{\sqrt{3}}{2}\right) - \left(\frac{\sqrt{2}}{2}\right)\left(\frac{1}{2}\right)$$

$$= -\frac{\sqrt{6} + \sqrt{2}}{4}.$$

24. Since $195° = 135° + 60°$,

$$\cos 195° = \cos 135° \cos 60° - \sin 135° \sin 60° = \left(-\frac{\sqrt{2}}{2}\right)\left(\frac{1}{2}\right) - \left(\frac{\sqrt{2}}{2}\right)\left(\frac{\sqrt{3}}{2}\right)$$

$$= -\frac{\sqrt{2} + \sqrt{6}}{4}.$$

27. Since $345° = 360° - 15° = 360° - (45° - 30°) = 315° + 30°$,

$$\cos 345° = \cos 315° \cos 30° - \sin 315° \sin 30° = \left(\frac{\sqrt{2}}{2}\right)\left(\frac{\sqrt{3}}{2}\right) - \left(-\frac{\sqrt{2}}{2}\right)\left(\frac{1}{2}\right)$$

$$= \frac{\sqrt{6} + \sqrt{2}}{4}.$$

30. Since $17\pi/12 = (14\pi + 3\pi)/12 = 7\pi/6 + \pi/4$,

$$\tan\frac{17\pi}{12} = \frac{\tan(7\pi/6) + \tan(\pi/4)}{1 - \tan(7\pi/6)\tan(\pi/4)} = \frac{1/\sqrt{3} + 1}{1 - (1/\sqrt{3})(1)} = \frac{1/\sqrt{3} + 1}{1 - 1/\sqrt{3}}\left(\frac{\sqrt{3}}{\sqrt{3}}\right)$$

$$= \frac{\sqrt{3} + 1}{\sqrt{3} - 1}.$$

33. We first note that

$$\cos 2x = \cos^2 x - \sin^2 x = (1 - \sin^2 x) - \sin^2 x = 1 - 2\sin^2 x.$$

Thus, letting $x = \pi/5$, we have

$$1 - 2\sin^2\frac{\pi}{5} = \cos\frac{2\pi}{5}.$$

36. Since $3\pi/2 < x < 2\pi$, x is in the fourth quadrant, so $\sin x$ is negative. Using $\sin^2 x + \cos^2 x = 1$ we have

$$\sin x = -\sqrt{1 - \cos^2 x} = -\sqrt{1 - \frac{3}{25}} = -\frac{\sqrt{22}}{5}.$$

(a) From the double-angle formula (14) in the text,

$$\cos 2x = \cos^2 x - \sin^2 x = \left(\frac{\sqrt{3}}{5}\right)^2 - \left(-\frac{\sqrt{22}}{5}\right)^2 = \frac{3}{25} - \frac{22}{25} = -\frac{19}{25}.$$

(b) From the double-angle formula (15) in the text,

$$\sin 2x = 2\sin x \cos x = 2\left(-\frac{\sqrt{22}}{5}\right)\left(\frac{\sqrt{3}}{5}\right) = -\frac{2\sqrt{66}}{25}.$$

(c) Since $\tan 2x = \sin 2x / \cos 2x$, we have

$$\tan 2x = \frac{-2\sqrt{66}/25}{-19/25} = \frac{2}{19}\sqrt{66}.$$

39. Since $\pi/2 < x < \pi$, x is in the second quadrant and $\sin x$ is positive. Using $\cos x = 1/\sec x$ we have

$$\cos x = \frac{1}{-13/5} = -\frac{5}{13},$$

and, using $\sin^2 x + \cos^2 x = 1$, we have

$$\sin x = \sqrt{1 - \cos^2 x} = \sqrt{1 - \left(-\frac{5}{13}\right)^2} = \frac{1}{13}\sqrt{169 - 25} = \frac{12}{13}.$$

(a) From the double-angle formula (14) in the text,

$$\cos 2x = \cos^2 x - \sin^2 x = \left(-\frac{5}{13}\right)^2 - \left(\frac{12}{13}\right)^2 = -\frac{119}{169}.$$

(b) From the double-angle formula (15) in the text,

$$\sin 2x = 2\sin x \cos x = 2\left(\frac{12}{13}\right)\left(-\frac{5}{13}\right) = -\frac{120}{169}.$$

(c) Since $\tan 2x = \sin 2x / \cos 2x$, we have

$$\tan 2x = \frac{-120/169}{-119/169} = \frac{120}{119}.$$

42. If we let $x = \pi/4$, then $x/2 = \pi/8$ and formula (20) in the text yields

$$\sin^2 \frac{\pi}{8} = \frac{1}{2}\left(1 - \cos\frac{\pi}{4}\right) = \frac{1}{2}\left(1 - \frac{\sqrt{2}}{2}\right) = \frac{2 - \sqrt{2}}{4}.$$

Since $\pi/8$ is in the first quadrant, $\sin(\pi/8) > 0$ and $\sin(\pi/8) = \frac{1}{2}\sqrt{2 - \sqrt{2}}$.

45. If we let $\theta = 135°$, then $\theta/2 = 67.5°$ and formula (19) in the text yields

$$\cos^2 67.5° = \frac{1}{2}(1 + \cos 135°) = \frac{1}{2}\left[1 + \left(-\frac{\sqrt{2}}{2}\right)\right] = \frac{2 - \sqrt{2}}{4}.$$

Since $67.5°$ is in the first quadrant, $\cos 67.5° > 0$ and $\cos 67.5° = \frac{1}{2}\sqrt{2 - \sqrt{2}}$.

48. We first note that since $\sec x$ is an even function, $\sec(-3\pi/8) = \sec(3\pi/8)$. If we let $x = 3\pi/4$, then $x/2 = 3\pi/8$ and formula (19) in the text yields

$$\cos^2(3\pi/8) = \frac{1}{2}[1 + \cos(3\pi/4)] = \frac{1}{2}\left[1 + \left(-\frac{\sqrt{2}}{2}\right)\right] = \frac{2 - \sqrt{2}}{4}.$$

Since $3\pi/8$ is in the first quadrant, $\cos(3\pi/8) > 0$ and $\cos(3\pi/8) = \frac{1}{2}\sqrt{2 - \sqrt{2}}$. Finally, $\sec x = 1/\cos x$, so

$$\sec\frac{3\pi}{8} = \frac{1}{\cos(3\pi/8)} = \frac{1}{\frac{1}{2}\sqrt{2 - \sqrt{2}}} = \frac{2}{\sqrt{2 - \sqrt{2}}}.$$

51. Since $\sec^2 x = 1 + \tan^2 x$ and $\tan x = 2$, we have $\sec^2 x = 1 + 4 = 5$ and $\cos^2 x = \frac{1}{5}$. Because $\pi < x < 3\pi/2$, x is in the third quadrant and $\cos x < 0$, so $\cos x = -1/\sqrt{5}$. Using formulas (19) and (20) in the text, we have

$$\cos^2\frac{x}{2} = \frac{1}{2}(1 + \cos x) = \frac{1}{2}\left(1 - \frac{1}{\sqrt{5}}\right) = \frac{\sqrt{5} - 1}{2\sqrt{5}} = \frac{5 - \sqrt{5}}{10}$$

and

$$\sin^2\frac{x}{2} = \frac{1}{2}(1 - \cos x) = \frac{1}{2}\left(1 + \frac{1}{\sqrt{5}}\right) = \frac{\sqrt{5} + 1}{2\sqrt{5}} = \frac{5 + \sqrt{5}}{10}.$$

Then, since $\pi < x < 3\pi/2$, we have $\pi/2 < x/2 < 3\pi/4$, so $x/2$ is in the second quadrant where $\cos x/2 < 0$ and $\sin x/2 > 0$. Thus,

$$\cos\frac{x}{2} = -\sqrt{\frac{5 - \sqrt{5}}{10}}, \qquad \sin\frac{x}{2} = \sqrt{\frac{5 + \sqrt{5}}{10}},$$

and

$$\tan\frac{x}{2} = \frac{\sin(x/2)}{\cos(x/2)} = \frac{\sqrt{(5+\sqrt{5})/10}}{-\sqrt{(5-\sqrt{5})/10}} = -\sqrt{\frac{5+\sqrt{5}}{5-\sqrt{5}}} = -\sqrt{\frac{3+\sqrt{5}}{2}}.$$

(**Note:** The value of $\tan(x/2)$ above does not appear to be the same as that given in the answer section of the text, but it can be shown that the two values are equivalent. To get a sense of this, you could simply use a calculator to find the decimal equivalent of each.)

54. From $\cot x = -\frac{1}{4}$ we infer that $\tan x = 1/\cot x = -4$. Since $\sec^2 x = 1 + \tan^2 x$ and $\tan x = -4$, we have $\sec^2 x = 1 + 16 = 17$ so $\cos^2 x = \frac{1}{17}$. Because $90° < x < 180°$, x is in the second quadrant and $\cos x < 0$, so $\cos x = -1/\sqrt{17}$. Using formulas (19) and (20) in the text, we have

$$\cos^2\frac{x}{2} = \frac{1}{2}(1 + \cos x) = \frac{1}{2}\left(1 - \frac{1}{\sqrt{17}}\right) = \frac{\sqrt{17}-1}{2\sqrt{17}} = \frac{17-\sqrt{17}}{34},$$

and

$$\sin^2\frac{x}{2} = \frac{1}{2}(1 - \cos x) = \frac{1}{2}\left(1 + \frac{1}{\sqrt{17}}\right) = \frac{\sqrt{17}+1}{2\sqrt{17}} = \frac{17+\sqrt{17}}{34}.$$

Now, since $90° < x < 180°$, we have $45° < x/2 < 90°$, so $x/2$ is in the first quadrant where both $\cos(x/2)$ and $\sin(x/2)$ are positive. Thus,

$$\cos\frac{x}{2} = \sqrt{\frac{17-\sqrt{17}}{34}}, \qquad \sin\frac{x}{2} = \sqrt{\frac{17+\sqrt{17}}{34}},$$

and

$$\tan\frac{x}{2} = \frac{\sin(x/2)}{\cos(x/2)} = \frac{\sqrt{(17+\sqrt{17})/34}}{\sqrt{(17-\sqrt{17})/34}} = \sqrt{\frac{17-\sqrt{17}}{17+\sqrt{17}}} = \sqrt{\frac{9+\sqrt{17}}{8}}.$$

57. For $y = \cos\pi x - \sin\pi x$ we identify, in (23) in the text, $c_1 = 1$, $c_2 = -1$, and $B = \pi$. Then, using (24) and (25) in the text, we have

$$A = \sqrt{c_1^2 + c_2^2} = \sqrt{1^2 + (-1)^2} = \sqrt{2}$$

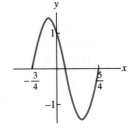

and

$$\sin\phi = \frac{1}{\sqrt{2}}, \qquad \cos\phi = -\frac{1}{\sqrt{2}}.$$

From the last two equations, we conclude that ϕ can be taken to be the second quadrant angle $3\pi/4$. Therefore,

$$y = \cos\pi x - \sin\pi x = \sqrt{2}\sin\left(\pi x + \frac{3\pi}{4}\right).$$

The amplitude is $\sqrt{2}$, the period is $2\pi/\pi = 2$, and the phase shift is $\frac{3}{4}$ units to the left.

60. For $y = \sqrt{3}\cos 4x - \sin 4x$ we identify, in (23) in the text, $c_1 = \sqrt{3}$, $c_2 = -1$, and $B = 4$. Then, using (24) and (25) in the text, we have

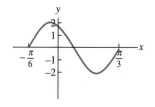

$$A = \sqrt{c_1^2 + c_2^2} = \sqrt{(\sqrt{3})^2 + (-1)^2} = \sqrt{3+1} = 2$$

and

$$\sin\phi = \frac{\sqrt{3}}{2}, \qquad \cos\phi = -\frac{1}{2}.$$

From the last two equations, we conclude that ϕ can be taken to be the second quadrant angle $2\pi/3$. Therefore,

$$y = \sqrt{3}\cos 4x - \sin 4x = 2\sin(4x + 2\pi/3).$$

The amplitude is 2, the period is $2\pi/4 = \pi/2$, and the phase shift is $(2\pi/3)/4 = \pi/6$ units to the left.

63. We want to solve $\sin[(\pi/6)/2] = \sin(\pi/12) = 1/M$. From (20) in the text we have

$$\sin^2\frac{\pi}{12} = \frac{1}{2}\left(1 - \cos\frac{\pi}{6}\right) = \frac{1}{2}\left(1 - \frac{\sqrt{3}}{2}\right) = \frac{2-\sqrt{3}}{4}.$$

Since $\pi/12$ is in the first quadrant, we use the positive square root when computing

$$\sin\frac{\pi}{12} = \sqrt{\frac{2-\sqrt{3}}{4}} = \frac{1}{2}\sqrt{2-\sqrt{3}}.$$

Thus,

$$M = \frac{1}{\sin(\pi/12)} = \frac{2}{\sqrt{2-\sqrt{3}}} \approx 3.86.$$

66. From $\cos 2\phi = gh/(v_0^2 + gh)$ we have

$$\cos^2\phi = \frac{1}{2}(1 + \cos 2\phi) = \frac{1}{2}\left(1 + \frac{gh}{v_0^2 + gh}\right) = \frac{1}{2}\frac{v_0 + 2gh}{v_0^2 + gh}$$

and

$$\sin^2\phi = \frac{1}{2}(1 - \cos 2\phi) = \frac{1}{2}\left(1 - \frac{gh}{v_0^2 + gh}\right) = \frac{1}{2}\frac{v_0}{v_0^2 + gh}.$$

Then

$$R_{\max} = \frac{v_0^2 \cos\phi}{g} \left(\sin\phi + \sqrt{\sin^2\phi + \frac{2gh}{v_0^2}} \right)$$

$$= \frac{v_0^2}{g} \frac{1}{\sqrt{2}} \sqrt{\frac{v_0^2 + 2gh}{v_0^2 + gh}} \left(\frac{1}{\sqrt{2}} \sqrt{\frac{v_0^2}{v_0^2 + gh}} + \sqrt{\frac{1}{2}\left(\frac{v_0^2}{v_0^2 + gh}\right) + \frac{2gh}{v_0^2}} \right)$$

$$= \frac{v_0^2}{2g} \sqrt{\frac{v_0^2 + 2gh}{v_0^2 + gh}} \left(\frac{v_0}{\sqrt{v_0^2 + gh}} + \sqrt{\frac{v_0^4 + 4gh(v_0^2 + gh)}{v_0^2(v_0^2 + gh)}} \right)$$

$$= \frac{v_0^2 \sqrt{v_0^2 + 2gh} \left(v_0 + \sqrt{v_0^4 + 4v_0^2 gh + 4g^2h^2}/v_0 \right)}{2g(v_0^2 + gh)}$$

$$= \frac{v_0 \sqrt{v_0^2 + 2gh} \left(v_0^2 + \sqrt{(v_0^2 + 2gh)^2} \right)}{2g(v_0^2 + gh)}$$

$$= \frac{v_0 \sqrt{v_0^2 + 2gh} \, (2v_0^2 + 2gh)}{2g(v_0^2 + gh)}$$

$$= \frac{v_0 \sqrt{v_0^2 + 2gh}}{g}.$$

4.6   Trigonometric Equations

3. If $\sec x = \sqrt{2}$, the reference angle for x is $\pi/4$. Since the value of $\sec x$ is positive, the terminal side of the angle lies in either the first or fourth quadrant. Thus, as shown in the figure, the only solutions between 0 and 2π are $x = \pi/4$ or $x = 7\pi/4$. Since the secant function is 2π-periodic, all of the remaining solutions can be obtained by adding integer multiples of 2π to these solutions:

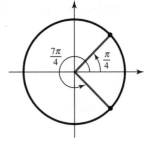

$$x = \frac{\pi}{4} + 2n\pi \qquad \text{or} \qquad x = \frac{7\pi}{4} + 2n\pi,$$

where n is an integer.

6. If $\csc x = 2$, the reference angle for x is $\pi/6$. Since the value of $\csc x$ is positive, the terminal side of the angle lies in either the first or second quadrant. Thus, as shown in the figure, the only solutions between 0 and 2π are $x = \pi/6$ or $x = 5\pi/6$. Since the cosecant function is 2π-periodic, all of the remaining solutions can be obtained by adding integer multiples of 2π to these solutions:

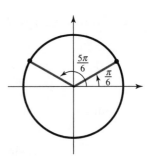

$$x = \frac{\pi}{6} + 2n\pi \qquad \text{or} \qquad x = \frac{5\pi}{6} + 2n\pi,$$

where n is an integer.

9. Since $\tan x = 0$ for $x = 0$ and $x = \pi$, these are the only solutions of $\tan x = 0$ in $[0, 2\pi)$. The tangent function is π-periodic, so all of the remaining solutions can be obtained by adding integer multiples of π to $x = 0$. Thus, all solutions are $x = n\pi$, where n is an integer.

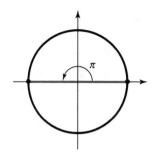

12. If $\sqrt{3}\cot x = 1$, then $\cot x = 1/\sqrt{3}$ and the reference angle for x is $\pi/3$. Since the value of $\cot x$ is positive, the terminal side of the angle lies in either the first or third quadrant. Thus, as shown in the figure, the only solutions between 0 and 2π are $x = \pi/3$ or $x = 4\pi/3$. Since the cotangent function is π-periodic, all of the remaining solutions can be obtained by adding integer multiples of π to these solutions: $x = \frac{\pi}{3} + n\pi$, where n is an integer.

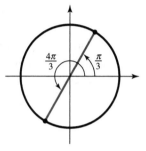

15. If $1 + \cot\theta = 0$, then $\cot\theta = -1$ and the reference angle for θ is $135°$. Since the value of $\cot\theta$ is negative, the terminal side of the angle lies in either the second or fourth quadrant. Thus, as shown in the figure, the only solutions between $0°$ and $360°$ are $\theta = 135°$ or $\theta = 315°$. Since the cotangent function has period $180°$, all of the remaining solutions can be obtained by adding integer multiples of $180°$ to these solutions: $\theta = 135° + 180°n$, where n is an integer.

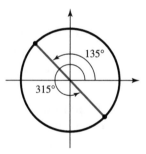

18. If $2\cos\theta + \sqrt{2} = 0$, then $\cos\theta = -\sqrt{2}/2$ and the reference angle for θ is $135°$. Since the value of $\cos\theta$ is negative, the terminal side of the angle lies in either the second or third quadrant. Thus, as shown in the figure, the only solutions between $0°$ and $360°$ are $\theta = 135°$ or $\theta = 225°$. Since the cosine function has period $360°$, all of the remaining solutions can be obtained by adding integer multiples of $360°$ to these solutions:

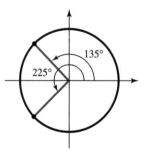

$$x = 135° + 360°n \qquad \text{or} \qquad x = 225° + 360°n,$$

where n is an integer.

21. Since $\sec x \geq 1$ for all x, we can divide the equation by $\sec x$, obtaining $3\sec x = 1$ or $\sec x = \frac{1}{3}$. But, again, since $\sec x \geq 1$ for all x, this equation has no solutions. Thus, the original equation, $3\sec^2 x = \sec x$, has no solutions.

24. Factoring the equation, we have $(2\sin\theta + 1)(\sin\theta - 1) = 0$, from which we conclude that $\sin\theta = -\frac{1}{2}$ and $\sin\theta = 1$. We first consider $\sin\theta = -\frac{1}{2}$. The reference angle for θ is $30°$. Since the value of $\sin\theta$ is negative, the terminal side of the angle lies in either the third or fourth quadrants. Thus, as shown in the figure, the only solutions between $0°$ and $360°$ are $\theta = 210°$ or $\theta = 330°$. Since the sine function has period $360°$, all of the remaining solutions can be obtained by adding integer multiples of $360°$ to these solutions:

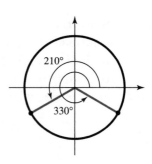

$$\theta = 210° + 360°n \qquad \text{or} \qquad \theta = 330° + 360°n,$$

where n is an integer. Since $\sin\theta = 1$ for $0° \le \theta \le 360°$ only when $\theta = 90°$, the solutions of $\sin\theta = 1$ are $90° + 360°n$, where n is an integer. Therefore, all solutions of $2\sin^2\theta - \sin\theta - 1 = 0$ are
$$\theta = 210° + 360°n, \qquad 330° + 360°n, \qquad \text{or} \qquad \theta = 90° + 360°n,$$

where n is an integer.

27. Since $\cos 2x = -1$ on $[0, 2\pi)$ only when $2x = \pi$, the solutions of $\cos 2x = -1$ are $2x = \pi + 2n\pi$ or $x = \pi/2 + n\pi$, where n is an integer.

30. If $\tan 4\theta = -1$, the reference angle for 4θ is $135°$. Since the value of $\tan 4\theta$ is negative, the terminal side of the angle lies in either the second or fourth quadrant. Thus, as shown in the figure with angles 4θ, the only solutions between $0°$ and $360°$ are $\theta = 135°$ or $\theta = 315°$. Since the tangent function has period $180°$, all of the remaining solutions can be obtained by adding integer multiples of $180°$ to these solutions:

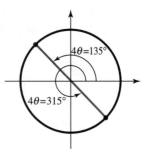

$$4\theta = 135° + 180°n \qquad \text{or} \qquad \theta = 33.75° + 45°n,$$

where n is an integer.

33. Using the double-angle formula and then factoring, we have

$$\sin 2x + \sin x = 0$$
$$2\sin x \cos x + \sin x = 0$$
$$\sin x(2\cos x + 1) = 0.$$

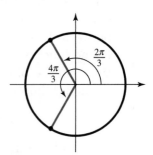

Thus, $\sin x = 0$, so $x = nx$, where n is an integer, and $\cos x = -\frac{1}{2}$. In this case, the reference angle for x is $\pi/3$. Since the value of $\cos x$ is negative, the terminal side of the angle lies in either the second or third quadrant. Thus, as shown in the figure, the only solutions between 0 and 2π are $x = 2\pi/3$ or $x = 4\pi/3$. Since the cosine function is 2π-periodic, all of the remaining solutions can be obtained by adding integer multiples of 2π to these solutions:

$$x = \frac{2\pi}{3} + 2n\pi \qquad \text{or} \qquad x = \frac{4\pi}{3} + 2n\pi,$$

where n is an integer. Therefore, all solutions of $\sin 2x + \sin x = 0$ are

$$x = n\pi, \quad x = \frac{2\pi}{3} + 2n\pi, \quad \text{or} \quad x = \frac{4\pi}{3} + 2n\pi,$$

where n is an integer.

36. Using the double-angle formula and then factoring, we have

$$\sin 2\theta + 2\sin\theta - 2\cos\theta = 2$$
$$2\sin\theta\cos\theta + 2\sin\theta - 2\cos\theta - 2 = 0$$
$$2\sin\theta(\cos\theta + 1) - 2(\cos\theta + 1) = 0$$
$$2(\sin\theta - 1)(\cos\theta + 1) = 0.$$

Thus, $\sin\theta = 1$ and $\theta = 90° + 360°n$ or $\cos\theta = -1$ and $\theta = 180° + 360°n$, where n is an integer.

39. Using $\sec x = 1/\cos x$ and $\tan x = \sin x/\cos x$ we have

$$\sec x \sin^2 x = \tan x$$
$$\frac{\sin^2 x}{\cos x} = \frac{\sin x}{\cos x}$$
$$\sin^2 x = \sin x$$
$$\sin^2 x - \sin x = 0$$
$$\sin x(\sin x - 1) = 0.$$

We note that $\cos x \neq 0$, and hence $x \neq \pi/2 + n\pi$ where n is an integer, because the secant and tangent functions are not defined when $\cos x = 0$. Then $\sin x = 0$, so $x = n\pi$, and $\sin x = 1$, so $x = \pi/2 + 2n\pi$, where n is an integer. We have already seen that $x = \pi/2 + 2n\pi$ cannot be a solution, so the only solutions are $x = n\pi$, where n is an integer.

42. We rewrite $\sin x + \cos x = 0$ as $\sin x = -\cos x$. Since $\cos x$ and $\sin x$ are never 0 for the same value of x, we conclude that $\cos x \neq 0$ for any values of x that are solutions of the equation. Then $\sin x + \cos x = 0$ is equivalent to $\sin x/\cos x = \tan x = -1$. Thus, $x = 3\pi/4$ and, since the tangent function is π-periodic, solutions are $x = \frac{3\pi}{4} + n\pi$, where n is an integer.

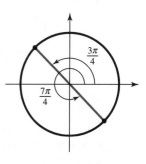

45. We begin by writing the equation in the form $\cos\theta = \sqrt{\cos\theta}$ and squaring both sides:

$$\cos\theta = \sqrt{\cos\theta}$$
$$\cos^2\theta = \cos\theta$$
$$\cos^2\theta - \cos\theta = 0$$
$$\cos\theta(\cos\theta - 1) = 0.$$

This implies $\cos\theta = 0$, so $\theta = 90° + 180°n$, and $\cos\theta = 1$, so $\theta = 360°n$, where n is an integer. Since we squared both sides of the equation, it is imperative that we check the answers. When $\theta = 90° + 180°n$,

$$\cos(90° + 180°n) = 0 = \sqrt{\cos(90° + 180°n)};$$

and when $\theta = 360°n$,

$$\cos(360°n) = 1 = \sqrt{\cos(360°n)}.$$

The solutions of $\cos\theta - \sqrt{\cos\theta} = 0$ are

$$\theta = 90° + 180°n \qquad \text{or} \qquad \theta = 360°n,$$

where n is an integer.

48. Solving $f(x) = 2\cos(x + \pi/4) = 0$ we obtain $x + \pi/4 = \pi/2 + n\pi$, or $x = \pi/4 + n\pi$, where n is an integer. The first three positive x-intercepts are at $\pi/4$, $\pi/4 + \pi$, and $\pi/4 + 2\pi$, so the intercepts are $(\pi/4, 0)$, $(5\pi/4, 0)$, and $(9\pi/4, 0)$.

51. We need to solve $f(x) = \sin x + \tan x = 0$. Writing $\tan x = \sin x / \cos x$ and factoring, we have

$$\sin x + \frac{\sin x}{\cos x} = 0$$

$$\sin x \left(1 + \frac{1}{\cos x}\right) = 0$$

so $\sin x = 0$ and $\cos x = -1$. Thus, solutions are $x = n\pi$ or $x = (2n + 1)\pi$, where n is an integer. That is, the set of solutions is $x = n\pi$, where n is an integer. The first three positive x-intercepts are $(\pi, 0)$, $(2\pi, 0)$, and $(3\pi, 0)$.

54. Using $\cos 3x = \cos(x + 2x) = \cos x \cos 2x - \sin x \sin 2x$, we have

$$\begin{aligned}
f(x) &= \cos x + \cos 3x = \cos x + \cos x \cos 2x - \sin x \sin 2x \\
&= \cos x(1 + \cos 2x) - \sin x(2\sin x \cos x) \\
&= \cos x(1 + \cos 2x - 2\sin^2 x) = \cos[1 + \cos 2x - 2(1 - \cos^3 x)] \\
&= \cos x(\cos 2x + 2\cos^2 x - 1) = \cos x(2\cos 2x) = 2\cos x \cos 2x.
\end{aligned}$$

Thus, $\cos x = 0$, so $x = \pi/2 + n\pi$, or $\cos 2x = 0$, so $2x = \pi/2 + m\pi$ and $x = \pi/4 + m\pi/2$, where n and m are integers. This means that the first three positive zeros of $f(x)$ are $\pi/4$, $\pi/2$, and $3\pi/4$. Here we use $m = 0$, $n = 0$, and $m = 1$, respectively.

57. Since $\cot x - x = 0$ is equivalent to $\cot x = x$, we see from the graphs of $y = \cot x$ and $y = x$ that there are infinitely many solutions of $\cot x - x = 0$.

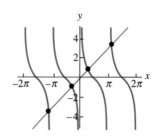

60. We need to determine when $y = 8\cos(\pi t - \pi/12) = 0$. This will occur when

$$\pi t - \pi/12 = \pi t - \frac{\pi}{12} = \frac{\pi}{2} + n\pi \qquad \text{or} \qquad t = \frac{7}{12} + n \text{ seconds,}$$

where n is a nonnegative integer (assuming t must be positive).

63. We need to solve $V = 110\sin(120\pi t - \pi/6) = 0$. This will occur when

$$120\pi t - \frac{\pi}{6} = n\pi \qquad \text{or} \qquad t = \frac{1}{720} + \frac{n}{120} \text{ seconds,}$$

where n is a nonnegative integer (assuming t must be positive).

4.7 Inverse Trigonometric Functions

3. Letting $y = \arccos(-1)$, we have that $\cos y = -1$, and we must find y such that $0 \le y \le \pi$. Since $\cos \pi = -1$, $y = \arccos(-1) = \pi$.

6. Letting $y = \arctan(-\sqrt{3})$, we have that $\tan y = -\sqrt{3}$, and we must find y such that $-\pi/2 < y < \pi/2$. Since $\tan(-\pi/3) = -\sqrt{3}$, $y = \arctan(-\sqrt{3}) = -\pi/3$.

9. Letting $y = \tan^{-1} 1$, we have that $\tan y = 1$, and we must find y such that $-\pi/2 < y < \pi/2$. Since $\tan(\pi/4) = 1$, $y = \tan^{-1} 1 = \pi/4$.

12. Letting $y = \arccos(-\frac{1}{2})$, we have that $\cos y = -\frac{1}{2}$, and we must find y such that $0 \le y \le \pi$. Since $\cos(2\pi/3) = -\frac{1}{2}$, $y = \arccos(-\frac{1}{2}) = 2\pi/3$.

15. Letting $t = \cos^{-1}\frac{3}{5}$, we want to find $\sin t$. Using $\cos t = \frac{3}{5}$ and $\sin^2 t + \cos^2 t = 1$ we see that

$$\sin^2 t + \left(\frac{3}{5}\right)^2 = 1 \qquad \text{or} \qquad \sin t = \sqrt{1 - \frac{9}{25}} = \sqrt{\frac{16}{25}} = \frac{4}{5}.$$

We used the positive square root because $\cos^{-1}\frac{3}{5}$ is in the first quadrant, so $\sin(\cos^{-1}\frac{3}{5})$ is positive. Thus, $\sin(\cos^{-1}\frac{3}{5}) = \frac{4}{5}$.

18. Letting $t = \arctan\frac{1}{4}$, we want to find $\sin t$. Using $\tan t = \frac{1}{4}$ and $1 + \tan^2 t = \sec^2 t$ we see that

$$1 + \left(\frac{1}{4}\right)^2 = \sec^2 t \qquad \text{or} \qquad \sec^2 t = \frac{17}{16},$$

so $\cos^2 t = 1/(17/16) = \frac{16}{17}$. Using $\sin^2 t + \cos^2 t = 1$ we have

$$\sin^2 t + \frac{16}{17} = 1 \qquad \text{or} \qquad \sin t = \sqrt{1 - \frac{16}{17}} = \frac{1}{\sqrt{17}}.$$

We used the positive square root because $\arctan \frac{1}{4}$ is in the first quadrant, so $\sin(\arctan \frac{1}{4})$ is positive. Thus, $\sin(\arctan \frac{1}{4}) = 1/\sqrt{17}$.

21. We use

$$\csc\left(\sin^{-1} \frac{3}{5}\right) = \frac{1}{\sin(\sin^{-1}(3/5))}.$$

Since $-1 \le \frac{3}{5} \le 1$, $\sin(\sin^{-1} \frac{3}{5}) = \frac{3}{5}$ by (ii) of the PROPERTIES OF THE INVERSE TRIGONOMETRIC FUNCTIONS in this section of the text and $\csc(\sin^{-1} \frac{3}{5}) = \frac{1}{3/5} = \frac{5}{3}$.

24. Since $-1 \le -\frac{4}{5} \le 1$, $\cos(\cos^{-1} -\frac{4}{5}) = -\frac{4}{5}$ by (iv) of the PROPERTIES OF THE INVERSE TRIGONOMETRIC FUNCTIONS in this section of the text.

27. Since $-\pi/2 \le \pi/16 \le \pi/2$, $\arcsin(\sin(\pi/16)) = \pi/16$ by (i) of the PROPERTIES OF THE INVERSE TRIGONOMETRIC FUNCTIONS in this section of the text.

30. Since $-\pi/2 \le 5\pi/6 \le \pi/2$, $\sin^{-1}(\sin(5\pi/6)) = 5\pi/6$ by (i) of the PROPERTIES OF THE INVERSE TRIGONOMETRIC FUNCTIONS in this section of the text.

33. Letting $t = \tan^{-1} x$, we want to find $\sin t$. Using $\tan t = x$ and $1 + \tan^2 t = \sec^2 t$, we see that $\sec^2 t = 1 + x^2$, so $\cos^2 t = 1/(1 + x^2)$. Using $\sin^2 t + \cos^2 t = 1$, we have

$$\sin^2 t + \frac{1}{1+x^2} = 1 \qquad \text{or} \qquad \sin t = \sqrt{1 - \frac{1}{1+x^2}} = \sqrt{\frac{x^2}{1+x^2}} = \frac{x}{\sqrt{1+x^2}}.$$

We used x in the numerator rather than $|x|$ because the range of $\tan^{-1} t$ is $(-\pi/2, \pi/2)$, and the sine function is positive for arguments in the first quadrant but negative for arguments in the fourth quadrant.

36. Using $\sec t = 1/\cos t$, we have

$$\sec(\arccos x) = \frac{1}{\cos(\arccos x)} = \frac{1}{x}$$

by (iv) of the PROPERTIES OF THE INVERSE TRIGONOMETRIC FUNCTIONS in this section of the text.

39. Letting $t = \arctan x$, we want to find $\csc t$. Since $\tan t = x$, $\cot t = 1/x$. Using $1 + \cot^2 t = \csc^2 t$, we have

$$1 + \frac{1}{x^2} = \csc^2 t \qquad \text{or} \qquad \csc t = \frac{\sqrt{1+x^2}}{x}.$$

We used x in the denominator rather than $|x|$ because the range of $\arctan t$ is $(-\pi/2, \pi/2)$, and the cosecant function is positive for arguments in the first quadrant but negative for arguments in the fourth quadrant.

42.

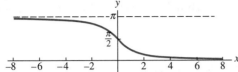

45.

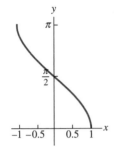

48. Letting $t = \arcsin x$, we want to find $\cos t$. Using $\sin t = x$ and $\sin^2 t + \cos^2 t = 1$, we have

$$x^2 + \cos^2 t = 1 \qquad \text{or} \qquad \cos t = \sqrt{1 - x^2}.$$

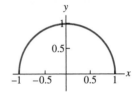

We used the positive square root because $-\pi/2 \le t \le \pi/2$ and the cosine of an angle in the first or fourth quadrant is positive. Thus, $\cos(\arcsin x) = \sqrt{1 - x^2}$, which is the upper half of the unit circle centered at the origin.

51. The equation $\tan^2 x + \tan x - 1 = 0$ is quadratic in $\tan x$. Since it does not factor with integer coefficients, we apply the quadratic formula to obtain

$$\tan x = \frac{-1 \pm \sqrt{1^2 - 4(1)(-1)}}{2(1)} = -\frac{1}{2} \pm \frac{1}{2}\sqrt{5}.$$

Using a calculator, we find

$$\tan^{-1}\left(-\frac{1}{2} - \frac{1}{2}\sqrt{5}\right) \approx -1.0172 \qquad \text{and} \qquad \tan^{-1}\left(-\frac{1}{2} + \frac{1}{2}\sqrt{5}\right) \approx 0.5536.$$

54. The equation $\tan^4 x - 3\tan^2 x + 1 = 0$ is quadratic in $\tan^2 x$. Since it does not factor with integer coefficients, we apply the quadratic formula to obtain

$$\tan^2 x = \frac{3 \pm \sqrt{(-3)^2 - 4(1)(1)}}{2(1)} = \frac{3}{2} \pm \frac{1}{2}\sqrt{5}.$$

Thus, $\tan x = \pm\sqrt{3/2 \pm \sqrt{5}/2}$, where we note that both $3/2 - \sqrt{5}/2$ and $3/2 + \sqrt{5}/2$ are positive. The solutions of the equation are

$$\tan^{-1}\sqrt{\frac{3}{2} + \frac{\sqrt{5}}{2}} \approx 0.5536$$

$$\tan^{-1}\left(-\sqrt{\frac{3}{2} - \frac{\sqrt{5}}{2}}\right) \approx -0.5536$$

$$\tan^{-1}\sqrt{\frac{3}{2} + \frac{\sqrt{5}}{2}} \approx 1.0172$$

$$\tan^{-1}\left(-\sqrt{\frac{3}{2} + \frac{\sqrt{5}}{2}}\right) \approx -1.0172.$$

57. The domain of $\operatorname{arccot} x$ is $(-\infty, \infty)$ and the range is $(0, \pi)$.

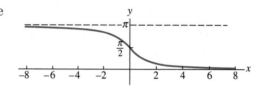

60. The graph of $\sec x$ restricted to the domain $[0, \pi/2) \cup [\pi, 3\pi/2)$ is shown as a black curve, while the graph of $\operatorname{arcsec} x$ is blue. The domain of $\operatorname{arcsec} x$ is $(-\infty, -1] \cup [1, \infty)$ and its range is $[0, \pi/2) \cup [\pi, 3\pi/2)$.

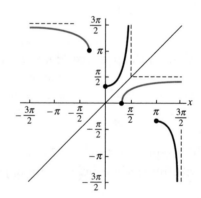

63. **(a)** $\sec(\operatorname{arcsec} x) = x$ if $-\infty < x \le -1$ or $1 \le x < \infty$.

   (b) $\operatorname{arcsec}(\sec x) = x$ if $0 \le x < \pi/2$ or $\pi/2 < x < \pi$.

   They are obtained simply by using the domain of $\operatorname{arcsec} x$ in part (a) and the domain of $\sec x$ in part (b).

66. Let $y = \operatorname{arcsec} x$. Then $x = \sec y = 1/\cos y$ and $\cos y = 1/x$. This implies that $y = \arccos(1/x)$. Thus, $\operatorname{arcsec} x = \arccos(1/x)$. The domains of both $\operatorname{arcsec} x$ and $\arccos(1/x)$ are both $|x| \ge 1$.

69. $\operatorname{arcsec}(-1.5) = \operatorname{arcsec}(-3/2) = \arccos(-1/(3/2)) = \arccos(-\frac{2}{3}) = 2.3005$

72. $\sec^{-1} 2.5 = \sec^{-1} \frac{5}{2} \cos^{-1}(1/(5/2)) = \cos^{-1} \frac{2}{5} = 1.1593$

75. Since R and g both are given in terms of feet, we convert 30 mph to feet per second using the facts that 1 mile = 5280 feet and 1 hour = 3600 seconds:

$$\frac{30(1 \text{ mile})}{1 \text{ hr}} = \frac{30(5280 \text{ feet})}{3600 \text{ seconds}} = 44 \text{ ft/s}.$$

Then

$$\tan\theta = \frac{(44 \text{ ft/s})^2}{(600 \text{ ft})(32 \text{ ft/s}^2)} = 0.1008333333,$$

and $\theta = \tan^{-1}(0.1008333333) = 0..1004936667$ radians $= 5.76°$.

78. Formula (7) in Section 4.5 is

$$\sin(x_1 + x_2) = \sin x_1 \cos x_2 + \cos x_1 \sin x_2.$$

If we identify $x_1 = \omega t + \theta$ and $x_2 = \phi$, then

$$\sin(\omega t + \theta)\cos\phi + \cos(\omega t + \theta)\sin\phi = \sin(\omega t + \theta + \phi)$$

and

$$I = I_0 \sin(\omega t + \theta + \phi).$$

Solving for t, we find

$$\sin(\omega t + \theta + \phi) = \frac{I}{I_0}$$

$$\omega t + \theta + \phi = \sin^{-1}\left(\frac{I}{I_0}\right)$$

$$t = \frac{1}{\omega}\left[\sin^{-1}\left(\frac{I}{I_0}\right) - \theta - \phi\right].$$

4.8   Right Triangle Trigonometry

3. The hypotenuse c is given by

$$c^2 = 2^2 + 6^2 = 40 \qquad \text{or} \qquad c = \sqrt{40} = 2\sqrt{10}.$$

Then

$$\sin\theta = \frac{\text{opp}}{\text{hyp}} = \frac{6}{2\sqrt{10}} = \frac{3}{\sqrt{10}} \qquad\qquad \cos\theta = \frac{\text{adj}}{\text{hyp}} = \frac{2}{2\sqrt{10}} = \frac{1}{\sqrt{10}}$$

$$\tan\theta = \frac{\text{opp}}{\text{adj}} = \frac{6}{2} = 3 \qquad\qquad \cot\theta = \frac{\text{adj}}{\text{opp}} = \frac{2}{6} = \frac{1}{3}$$

$$\sec\theta = \frac{\text{hyp}}{\text{adj}} = \frac{2\sqrt{10}}{2} = \sqrt{10} \qquad\qquad \csc\theta = \frac{\text{hyp}}{\text{opp}} = \frac{2\sqrt{10}}{6} = \frac{\sqrt{10}}{3}.$$

6. The side a opposite the angle θ is given by

$$a^2 + 1^2 = (\sqrt{5})^2 \qquad \text{or} \qquad a = \sqrt{5-1} = 2.$$

Then

$$\sin\theta = \frac{\text{opp}}{\text{hyp}} = \frac{2}{\sqrt{5}} \qquad\qquad \cos\theta = \frac{\text{adj}}{\text{hyp}} = \frac{1}{\sqrt{5}}$$

$$\tan\theta = \frac{\text{opp}}{\text{adj}} = \frac{2}{1} = 2 \qquad\qquad \cot\theta = \frac{\text{adj}}{\text{opp}} = \frac{1}{2}$$

$$\sec\theta = \frac{\text{hyp}}{\text{adj}} = \frac{\sqrt{5}}{1} = \sqrt{5} \qquad\qquad \csc\theta = \frac{\text{hyp}}{\text{opp}} = \frac{\sqrt{5}}{2}\,.$$

9. The hypotenuse c is given by

$$c^2 = x^2 + y^2 \qquad \text{or} \qquad c = \sqrt{x^2 + y^2}.$$

Then

$$\sin\theta = \frac{\text{opp}}{\text{hyp}} = \frac{y}{\sqrt{x^2+y^2}} \qquad\qquad \cos\theta = \frac{\text{adj}}{\text{hyp}} = \frac{x}{\sqrt{x^2+y^2}}$$

$$\tan\theta = \frac{\text{opp}}{\text{adj}} = \frac{y}{x} \qquad\qquad \cot\theta = \frac{\text{adj}}{\text{opp}} = \frac{x}{y}$$

$$\sec\theta = \frac{\text{hyp}}{\text{adj}} = \frac{\sqrt{x^2+y^2}}{x} \qquad\qquad \csc\theta = \frac{\text{hyp}}{\text{opp}} = \frac{\sqrt{x^2+y^2}}{y}\,.$$

12. Since $c = 10$ and $\beta = 49°$, we have

$$\sin\beta = \frac{b}{c} \qquad \text{so} \qquad b = c\sin\beta = 10\sin 49° \approx 7.55$$

and

$$\cos\beta = \frac{a}{c} \qquad \text{so} \qquad a = c\cos\beta = 10\cos 49° \approx 6.56.$$

15. Since $b = 1.5$ and $c = 3$, we have

$$\cos\alpha = \frac{b}{c} = \frac{1.5}{3} = \frac{1}{2} \qquad \text{so} \qquad \alpha = \cos^{-1}\frac{1}{2} = 60°.$$

Then $\beta = 90° - \alpha = 90° - 60° = 30°$, and

$$a^2 + b^2 = c^2 \qquad \text{so} \qquad a = \sqrt{c^2 - a^2} = \sqrt{3^2 - (1.5)^2} \approx 2.60.$$

18. Since $b = 4$ and $\alpha = 58°$, we have

$$\tan\alpha = \frac{a}{b} \qquad \text{so} \qquad a = b\tan\alpha = 4\tan 58° \approx 6.40$$

and

$$\cos\alpha = \frac{b}{c} \qquad \text{so} \qquad c = \frac{b}{\cos\alpha} = \frac{4}{\cos 58°} \approx 7.55.$$

21. Since $b = 20$ and $\alpha = 23°$, we have

$$\tan \alpha = \frac{a}{b} \qquad \text{so} \qquad a = b \tan \alpha = 20 \tan 23° \approx 8.49$$

and

$$\cos \alpha = \frac{b}{c} \qquad \text{so} \qquad c = \frac{b}{\cos \alpha} = \frac{20}{\cos 23°} \approx 21.73.$$

24. Let b denote the distance between the two trees. Then

$$\tan \beta = \frac{b}{100} \qquad \text{so} \qquad b = 100 \tan \beta = 100 \tan 29.7° \approx 57.4 \text{ ft.}$$

27. Let a denote the height of the first building, b denote the difference in heights of the two buildings, and c denote the distance between the two buildings. (See the figure to the right.) Then

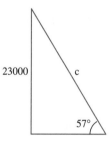

$$\tan 27° = \frac{150}{c} \qquad \text{so} \qquad c = \frac{150}{\tan 27°} \approx 294.3$$

and

$$\tan 41.42° = \frac{b}{c}$$

so

$$b = c \tan 41.42° = 294.3 \tan 41.42° \approx 259.72.$$

The height of the second building is

$$150 + b = 150 + 259.72 = 409.72 \text{ ft.}$$

30. The vertical distance between the radar station and the airplane is 23,000 ft. Let c denote the distance from the radar station to the airplane. Then, from the figure, we see that

$$\sin 57° = \frac{23,000}{c}$$

so

$$c = \frac{23,000}{\sin 57°} = \frac{23,000}{0.83868} \approx 27,404.36 \text{ ft.}$$

33. Let a be the distance from town A to the point C shown in the figure, and b the distance from town B to point C. Then

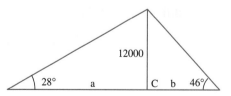

$$\tan 28° = \frac{12,000}{a}$$

and

$$\tan 46° = \frac{12,000}{b},$$

so the distance between towns A and B is

$$a + b = \frac{12,000}{\tan 28°} + \frac{12,000}{\tan 46°} = \frac{12,000}{0.5317} + \frac{12,000}{1.0355} \approx 34,156.98 \text{ ft} \approx 6.5 \text{ mi.}$$

36. Let h be the height of the head. Then, from the figure, we see that

$$\tan 79.946° = \frac{a}{1000} \quad \text{so} \quad a = 1000 \tan 79.946° \approx 5640.19,$$

and

$$\tan 80.05° = \frac{a+h}{1000} \quad \text{so} \quad a + h = 1000 \tan 80.05° \approx 5700.37.$$

The height of Washington's head is

$$h = 5700.37 - 5640.19 = 60.17 \text{ ft.}$$

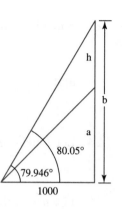

39. Denote the height of the storm by a, as indicated in the figure. Then

$$\tan 4° = \frac{a}{90} \quad \text{so} \quad a = 90 \tan 4° \approx 6.29 \text{ km.}$$

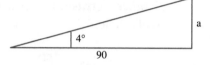

A plane capable of climbing to 10 km can easily fly over the storm.

42. (a) If the radius of the Earth is $R = 6400$, then its circumference at the equator is $C_e = 2\pi R \approx 40{,}212.39$ km. From $C_\theta = C_e \cos\theta$, when $\theta = 66.55°$, we see that the circumference of the Arctic Circle is

$$C_\theta = C_e \cos\theta \approx 40212.39 \cos 66.55° \approx 16{,}002.46 \text{ km.}$$

(b) When $\theta = 58.67°$ we have

$$C_\theta = C_e \cos\theta \approx 40212.39 \cos 58.67° \approx 20{,}909.09 \text{ km.}$$

45. Referring to the figure in the text, we let C be the point directly below the balloon and on the line through S_1 and S_2. Also, let A be the distance between S_1 and C and B the distance between C and S_2. Then

$$\tan\alpha = \frac{h}{a} \quad \text{so} \quad a = \frac{h}{\tan\alpha} = h\cot\alpha$$

and

$$\tan\beta = \frac{h}{b} \quad \text{so} \quad b = \frac{h}{\tan\alpha} = h\cot\beta.$$

Adding, we obtain

$$c = a + b = h\cos\alpha + h\cot\beta = h(\cot\alpha + \cot\beta),$$

so

$$h = \frac{c}{\cot\alpha + \cot\beta}.$$

48. Referring to the figure in the text, we see that

$$\cos\theta = \frac{1/2}{d} \quad \text{so} \quad d = \frac{1/2}{\cos\theta} = \frac{1}{2}\sec\theta.$$

4.9   Law of Sines and Law of Cosines

Note: In this section all answers are shown with two decimal place accuracy, but the computations were done on a calculator with all results stored in the calculator's memory. You will see somewhat different answers if you write down a rounded version of each result and then later key the rounded version back in.

3. With $\beta = 37°$ and $\gamma = 51°$ it follows that $\alpha = 180° - 37° - 51° = 92°$. Using the Law of Sines, we have

$$\frac{\sin \alpha}{a} = \frac{\sin \beta}{b} \qquad \text{or} \qquad \frac{\sin 92°}{5} = \frac{\sin 37°}{b}$$

so

$$b = \frac{5 \sin 37°}{\sin 92°} \approx 3.01.$$

Also,

$$\frac{\sin \alpha}{a} = \frac{\sin \gamma}{c} \qquad \text{or} \qquad \frac{\sin 92°}{5} = \frac{\sin 51°}{c}$$

so

$$c = \frac{5 \sin 51°}{\sin 92°} \approx 3.89.$$

6. With $\alpha = 120°$, $a = 9$, and $c = 4$, and using the Law of Sines, we have

$$\frac{\sin \alpha}{a} = \frac{\sin \gamma}{c} \qquad \text{or} \qquad \frac{\sin 120°}{9} = \frac{\sin \gamma}{4},$$

so

$$\sin \gamma = \frac{4 \sin 120°}{9} = \frac{4(\sqrt{3}/2)}{9} = \frac{2\sqrt{3}}{9} \approx 0.38.$$

Then $\gamma = \sin^{-1}(0.38) \approx 22.64°$. Thus, $\beta = 180° - 120° - 22.64° = 37.36°$. It remains to find b. Again using the Law of Sines, we have

$$\frac{\sin \alpha}{a} = \frac{\sin \beta}{b} \qquad \text{or} \qquad \frac{\sin 120°}{9} = \frac{\sin 37.36°}{b},$$

so

$$b = \frac{9 \sin 37.36°}{\sin 120°} \approx 6.31.$$

9. With $\gamma = 15°$, $a = 8$, and $c = 5$, and using the Law of Sines, we have

$$\frac{\sin \alpha}{a} = \frac{\sin \gamma}{c} \qquad \text{or} \qquad \frac{\sin \alpha}{8} = \frac{\sin 15°}{5}$$

so

$$\sin \alpha = \frac{8 \sin 15°}{5} \approx 0.41.$$

In this problem, we are given two sides and an angle opposite one of the sides, so there are potentially two solutions. From $\sin \alpha = 0.41$ we have $\alpha = 24.46°$ in the first quadrant and $\alpha = 180° - 24.46° = 155.54°$ in the second quadrant. For $\alpha = 24.46$ we have $\beta = 180° - 24.46° - 15° = 140.54°$. To find b in this case, we again use the Law of Sines:

$$\frac{\sin \beta}{b} = \frac{\sin \gamma}{c} \qquad \text{or} \qquad \frac{\sin 140.54°}{b} = \frac{\sin 15°}{c},$$

so

$$b = \frac{5 \sin 140.54°}{\sin 15°} \approx 12.28.$$

Next, for $\alpha = 155.54°$ we have $\beta = 180° - 155.54° - 15° = 9.46°$. To find b in this case, we use the Law of Sines:

$$\frac{\sin \beta}{b} = \frac{\sin \gamma}{c} \qquad \text{or} \qquad \frac{\sin 9.46°}{b} = \frac{\sin 15°}{c},$$

so

$$b = \frac{5 \sin 9.46°}{\sin 15°} \approx 3.18.$$

12. With $\beta = 48°$, $a = 7$, and $c = 6$, and using the Law of Cosines, we have

$$b^2 = a^2 + c^2 - 2ac \cos \beta = 7^2 + 6^2 - 84 \cos 48° \approx 28.79,$$

so $b \approx 5.37$. We use the Law of Sines to find α:

$$\frac{\sin \alpha}{a} = \frac{\sin \beta}{b} \qquad \text{or} \qquad \sin \alpha = \frac{7 \sin 48°}{5.37} \approx 0.97,$$

so $\alpha = \sin^{-1}(0.97) \approx 75.80°$. Finally, $\gamma = 180° - 75.80° - 48° = 56.20°$.

15. With $\gamma = 97.33°$, $a = 3$, and $b = 6$, and using the Law of Cosines, we have

$$c^2 = a^2 + b^2 - 2ab \cos \gamma = 3^2 + 6^2 - 36 \cos 97.33° \approx 49.59,$$

so $c = 7.04$. We use the Law of Sines to find α:

$$\frac{\sin \alpha}{a} = \frac{\sin \gamma}{c} \qquad \text{or} \qquad \sin \alpha = \frac{3 \sin 97.33°}{7.04} \approx 0.42,$$

so $\alpha = \sin^{-1}(0.42) \approx 24.99°$. Finally, $\beta = 180° - 24.99° - 97.33° \approx 57.68°$.

18. With $\alpha = 162°$, $b = 11$, and $c = 8$, and using the Law of Cosines, we have

$$a^2 = b^2 + c^2 - 2bc \cos \alpha = 11^2 + 8^2 - 176 \cos 162° \approx 352.39,$$

so $a = 18.77$. We use the Law of Sines to find β:

$$\frac{\sin \alpha}{a} = \frac{\sin \beta}{b} \qquad \text{or} \qquad \sin \beta = \frac{11 \sin 162°}{18.77} \approx 0.18,$$

so $\beta = \sin^{-1} 0.18 \approx 10.43°$. Finally, $\gamma = 180° - 162° - 10.43° = 7.57°$.

21. With $\gamma = 150°$, $b = 7$, and $c = 5$, and using the Law of Sines, we have

$$\frac{\sin\beta}{b} = \frac{\sin\gamma}{c} \qquad \text{or} \qquad \frac{\sin\beta}{7} = \frac{\sin 150°}{5} = \frac{1/2}{5} = \frac{1}{10},$$

so $\sin\beta = \frac{7}{10}$ and $\beta = \sin^{-1}(0.7) \approx 44.43$. Since $\beta + \gamma = 150 + 44.43 = 194.43 > 180°$, this is not possible, and there is no triangle with the given dimensions.

24. With $\alpha = 35°$, $a = 9$, and $b = 12$, and using the Law of Sines, we have

$$\frac{\sin\alpha}{a} = \frac{\sin\beta}{b} \qquad \text{or} \qquad \frac{\sin 35°}{9} = \frac{\sin\beta}{12},$$

so

$$\sin\beta = \frac{12\sin 35°}{9} \approx 0.76.$$

In this problem we are given two sides and an angle opposite one of the sides, so there are potentially two solutions. From $\sin\beta = 0.76$ we have $\beta = 49.89°$ in the first quadrant and $\beta = 180° - 49.89° = 130.11°$ in the second quadrant. For $\beta = 49.89°$ we have $\gamma = 180° - 35° - 49.89° = 95.11°$. To find c in this case we again use the Law of Sines:

$$\frac{\sin\alpha}{a} = \frac{\sin\gamma}{c} \qquad \text{or} \qquad \frac{\sin 35°}{9} = \frac{\sin 95.11°}{c},$$

so

$$c = \frac{9\sin 95.11°}{\sin 35°} \approx 15.63.$$

Next, for $\beta = 130.11°$ we have $\gamma = 180° - 35° - 130.11° = 14.89°$. To find c in this case we use the Law of Sines:

$$\frac{\sin\alpha}{a} = \frac{\sin\gamma}{c} \qquad \text{or} \qquad \frac{\sin 35°}{9} = \frac{\sin 14.89°}{c},$$

so

$$c = \frac{9\sin 14.89°}{\sin 35°} \approx 4.03.$$

27. With $a = 6$, $b = 8$, and $c = 12$, we see from the Law of Cosines that

$$a^2 = b^2 + c^2 - 2bc\cos\alpha \qquad \text{or} \qquad 6^2 = 8^2 + 12^2 - 192\cos\alpha$$

so $\cos\alpha = \frac{172}{192} = \frac{43}{48}$ and $\alpha = \cos^{-1}\frac{43}{48} \approx 26.38°$. To find β we can use a different version of the Law of Cosines:

$$b^2 = a^2 + c^2 - 2ac\cos\beta \qquad \text{or} \qquad 8^2 = 6^2 + 12^2 - 144\cos\beta,$$

so $\cos\beta = \frac{116}{144} = \frac{29}{36}$ and $\beta = \cos^{-1}\frac{29}{36} \approx 36.34°$. Finally,

$$\gamma = 180° - \alpha - \beta = 180° - 26.38° - 36.34° \approx 117.28°.$$

30. With $\alpha = 75°$, $\gamma = 45°$, and $b = 8$, it follows that $\beta = 180° - 75° - 45° = 60°$. Using the Law of Sines, we have

$$\frac{\sin\alpha}{a} = \frac{\sin\beta}{b} \qquad \text{or} \qquad \frac{\sin 75°}{a} = \frac{\sin 60°}{8}$$

so

$$a = \frac{8\sin 75°}{\sin 60°} \approx 8.92.$$

Also,

$$\frac{\sin\beta}{b} = \frac{\sin\gamma}{c} \qquad \text{or} \qquad \frac{\sin 60°}{8} = \frac{\sin 45°}{c}$$

so

$$c = \frac{8\sin 45°}{\sin 60°} \approx 6.53.$$

33. Using the figure in the text, we identify $\beta = 35°$, $\gamma = 115°$, and $b = 10$. We want to find the distance from A to B, which is the side c opposite $\gamma = 115°$. Using the Law of Sines, we have

$$\frac{\sin\beta}{b} = \frac{\sin\gamma}{c} \qquad \text{so} \qquad c = \frac{b\sin\gamma}{\sin\beta} = \frac{10\sin 115°}{\sin 35°} \approx 15.80 \text{ ft.}$$

36. Using the figure to the right, we note that $\beta = 90° - 31° = 59°$, and that the smaller angle in the upper, thinner triangle is $31° - \alpha$, as shown. Using the Law of Sines on this smaller triangle, we have

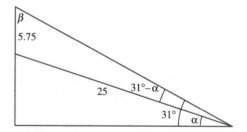

$$\frac{\sin(31° - \alpha)}{5.75} = \frac{\sin 59°}{25}$$

or

$$\sin(31° - \alpha) = \frac{5.75\sin 59°}{25} \approx 0.20.$$

Thus,

$$31° - \alpha = \sin^{-1}(0.20) \approx 11.37° \qquad \text{and} \qquad \alpha = 31° - 11.37° = 19.63°.$$

39. As shown in the figure to the right, the angle between the two headings is $62° + 90° = 152°$. From the Law of Cosines,

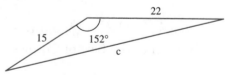

$$c^2 = 15^2 + 22^2 + 2(15)(22)\cos 152° \approx 1291.75$$

and $c \approx 35.94$ nautical miles.

42. After 3 hours, the first ship will travel $3(15) = 45$ nautical miles and the second ship will travel $3(12) = 35$ nautical miles. The angle between the paths of the two ships is $42° + 10° = 52°$. (See the figure to the right.)

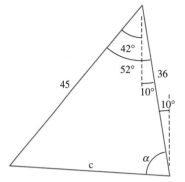

(a) We want to find the distance c between the two ships at the time the first ship runs aground. Using the Law of Cosines, we have

$$c^2 = 45^2 + 36^2 - 2(45)(36)\cos 52° \approx 1326.26,$$

so $c = 36.42$ nautical miles. If the ship is traveling at 14 knots it will take $36.42/14 \approx 2.60$ hours to reach the first ship.

(b) The bearing is $\alpha + 10°$ in the figure. From the Law of Sines, we have

$$\frac{\sin\alpha}{45} = \frac{\sin 52°}{c} \qquad \text{or} \qquad \sin\alpha = \frac{45\sin 52°}{36.42} \approx 0.97,$$

so

$$\alpha = \sin^{-1}(0.97) \approx 76.83°.$$

Thus, the bearing is $\alpha + 10° = 86.83°$ or N86.83°W.

4.10   Calculus Preview—The Limit Concept Revisited

3. Since the sine is an odd function, $\sin(-\theta) = -\sin\theta$ and

$$\lim_{\theta\to 0}\frac{\sin(-\theta)}{\theta} = \lim_{\theta\to 0}\left(-\frac{\sin\theta}{\theta}\right) = -\lim_{\theta\to 0}\frac{\sin\theta}{\theta} = -1.$$

6. Since the sine function is continuous at $x = \pi/4$,

$$\lim_{x\to\pi/4}\sin x = \sin\frac{\pi}{4} = \frac{1}{\sqrt 2}.$$

9. By (9) in the text,

$$\lim_{x\to 0}\frac{\cos x - 1}{10x} = \frac{1}{10}\lim_{x\to 0}\frac{\cos x - 1}{x} = \frac{1}{10}(0) = 0.$$

12. By (7) and (9) in the text,

$$\lim_{x\to 0}\frac{2\sin 4x + 1 - \cos x}{x} = 2\lim_{x\to 0}\frac{\sin 4x}{x} + \lim_{x\to 0}\frac{1 - \cos x}{x} = 2(4) + 0 = 8.$$

15. We use a trigonometric identity and (2) in the text:

$$\lim_{x\to\pi/2}\frac{\cos x}{\cot x} = \lim_{x\to\pi/2}\frac{\cos x}{\cos x/\sin x} = \lim_{x\to\pi/2}\sin x = \sin\frac{\pi}{2} = 1.$$

18. We use the double-angle formula for the cosine followed by (2) and (3) in the text:

$$\lim_{x \to \pi/4} \frac{\cos 2x}{\cos x - \sin x} = \lim_{x \to \pi/4} \frac{\cos^2 x - \sin^2 x}{\cos x - \sin x} = \lim_{x \to \pi/4} \frac{(\cos x + \sin x)(\cos x - \sin x)}{\cos x - \sin x}$$

$$= \lim_{x \to \pi/4} (\cos x + \sin x) = \cos \frac{\pi}{4} + \sin \frac{\pi}{4} = \frac{1}{\sqrt{2}} + \frac{1}{\sqrt{2}} = \frac{2}{\sqrt{2}}.$$

21. Since $f(\pi/6) = \sin(\pi/6) = \frac{1}{2}$, the point of tangency is $(\pi/6, 1/2)$. By (12) in the text, the derivative of $f(x) = \sin x$ is $f'(x) = \cos x$, so the slope of the tangent line at the point of tangency is

$$f'\left(\frac{\pi}{6}\right) = \cos \frac{\pi}{6} = \frac{\sqrt{3}}{2}.$$

Using the point-slope form of a line, an equation of the tangent line is

$$y - \frac{1}{2} = \frac{\sqrt{3}}{2}\left(x - \frac{\pi}{6}\right) \qquad \text{or} \qquad y = \frac{\sqrt{3}}{2}x - \frac{\sqrt{3}\pi}{12} + \frac{1}{2}.$$

24. Since $f(\pi/3) = \cos(\pi/3) = \frac{1}{2}$, the point of tangency is $(\pi/3, 1/2)$. By the result of Problem 23, the derivative of $f(x) = \cos x$ is $f'(x) = -\sin x$, so the slope of the tangent line at the point of tangency is

$$f'\left(\frac{\pi}{3}\right) = -\sin\left(\frac{\pi}{3}\right) = -\frac{\sqrt{3}}{2}.$$

Using the point-slope form of a line, an equation of the tangent line is

$$y - \frac{1}{2} = -\frac{\sqrt{3}}{2}\left(x - \frac{\pi}{3}\right) \qquad \text{or} \qquad y = -\frac{\sqrt{3}}{2}x + \frac{\sqrt{3}\pi}{6} + \frac{1}{2}.$$

Chapter 4 Review Exercises

3. The coordinates are $x = \cos t$ and $y = \sin t$. For $t = 5\pi/6$, $\cos(5\pi/6) = -\sqrt{3}/2$ and $\sin(5\pi/6) = \frac{1}{2}$. The point is $(-\sqrt{3}/2, 1/2)$.

6. Clearly $\pi < 8\pi/5 < 2\pi$, so the angle is in the third or fourth quadrant and we want to compare it to $3\pi/2$. Putting both numbers over a common denominator,

$$\frac{8\pi}{5} = \frac{16\pi}{10} \qquad \text{and} \qquad \frac{3\pi}{2} = \frac{15\pi}{10},$$

we see that $8\pi/5 > 3\pi/2$, so $8\pi/5$ is in the fourth quadrant.

9. The y-intercept occurs where $x = 0$, and

$$2\sec(0 + \pi) = 2\sec \pi = 2(-1) = -2,$$

so the y-intercept is $(0, -2)$.

12. Since $\cos^2(t/2) = \frac{1}{2}(1 + \cos t)$, for $\cos t = -\frac{2}{3}$ we have

$$\cos^2 \frac{t}{2} = \frac{1}{2}\left(1 - \frac{2}{3}\right) = \frac{1}{6},$$

so $\cos t/2 = 1/\sqrt{6}$ or $\cos t/2 = -1/\sqrt{6}$. Since t is in the third quadrant where $\cos t$ is negative, $\cos t/2 = -1/\sqrt{6}$.

15. Using the formula for the sum of the sine of two angles, we have

$$\sin\left(t + \frac{\pi}{4}\right) = \sin t \cos \frac{\pi}{4} + \cos t \sin \frac{\pi}{4} = (\sin t)\frac{1}{\sqrt{2}} + (\cos t)\frac{1}{\sqrt{2}} = \frac{1}{\sqrt{2}}(\sin t + \cos t).$$

Thus,

$$\sin t + \cos t = \sqrt{2}\sin\left(t + \frac{\pi}{4}\right).$$

18. Using the formula for the difference of the cosine of two angles, we have

$$\cos\left(\frac{\pi}{6} - \frac{5\pi}{4}\right) = \left(\cos \frac{\pi}{6}\right)\left(\cos \frac{5\pi}{4}\right) + \left(\sin \frac{\pi}{6}\right)\left(\sin \frac{5\pi}{4}\right)$$

$$= \frac{\sqrt{3}}{2}\left(-\frac{1}{\sqrt{2}}\right) + \frac{1}{2}\left(-\frac{1}{\sqrt{2}}\right) = -\frac{\sqrt{3}+1}{2\sqrt{2}}.$$

21. False; because $\tan t$ can be $\frac{3}{4}$, but neither $\sin t$ nor $\cos t$ can ever be larger than 1.

24. True; because $|\sec t| \geq 1$.

27. True; because $3\sin \pi(5) = 3\sin 5\pi = 3(0) = 0$.

30. True; because $\csc x$ is not defined for $x = 0$.

33. True; because $-1 \leq \sin x \leq 1$, so, multiplying by -2, we have $2 \geq -2\sin x \geq -2$ or $-2 \leq -2\sin x \leq 2$.

36. False; because the range of the arctangent function is $(-\pi/2, \pi/2)$ and $5\pi/4$ is not in this interval.

39. False; because

$$f(x + 2\pi) = (x + 2\pi)\sin(x + 2\pi) = (x + 2\pi)\sin x$$

$$= x\sin x + 2\pi \sin x = f(x) + 2\pi \sin x \neq f(x).$$

42. Writing the equation as $\cos t = \sin t$ we see that $1 = \sin t/\cos t = \tan t$. Thus, $t = \pi/4$ or $t = 5\pi/4$ in $[0, 2\pi]$ since $\tan t$ is positive in the first and third quadrants.

45. Squaring both sides of the equation, we have

$$(\sin t + \cos t)^2 = 1^2$$

$$\sin^2 t + 2\sin t\cos t + \cos^2 t = 1$$

$$2\sin t\cos t + (\sin^2 t + \cos^2 t) = 1$$

$$2\sin t\cos t + 1 = 1$$

$$2\sin t\cos t = 0.$$

The last equation is true when $\sin t = 0$ or $\cos t = 0$. Thus, solutions of $2 \sin t \cos t = 0$ are $t = 0$, π, 2π, $\pi/2$, or $3\pi/2$. Since squaring both sides of an equation can result in extraneous roots, it is essential that we check our answers. For $f(t) = \sin t + \cos t$, we find $f(0) = 1$, $f(\pi) = -1$, $f(2\pi) = 1$, $f(\pi/2) = 1$, and $f(3\pi/2) = -1$. Thus, solutions of $\sin t + \cos t = 1$ are $t = 0$, $t = 2\pi$, and $t = \pi/2$.

48. With $\gamma = 145°$, $a = 25$, and $c = 10$, and using the Law of Sines, we have

$$\frac{\sin \alpha}{a} = \frac{\sin \gamma}{c} \qquad \text{or} \qquad \frac{\sin \alpha}{25} = \frac{\sin 145°}{10},$$

so

$$\sin \alpha = \frac{25 \sin 145°}{10} \approx 1.43.$$

This is not possible because $\sin \alpha$ must be less than or equal to 1, so there is no triangle.

51. The range of the arccos function is $[0, \pi]$ and $\cos t$ is negative in the second quadrant, so $\cos^{-1}(-\frac{1}{2}) = \pi - \pi/3 = 2\pi/3$.

54. Letting $t = \arcsin \frac{2}{5}$, we want to find $\cos t$. Using $\sin t = \frac{2}{5}$ and $\sin^2 t + \cos^2 t = 1$ we see that

$$\left(\frac{2}{5}\right)^2 + \cos^2 t = 1 \qquad \text{or} \qquad \cos t = \sqrt{1 - \frac{4}{25}} = \sqrt{\frac{21}{25}} = \frac{\sqrt{21}}{5}.$$

We used the positive square root because $\arcsin \frac{2}{5}$ is in the first quadrant, so $\cos(\arcsin \frac{2}{5})$ is positive. Thus, $\cos(\arcsin \frac{2}{5}) = \sqrt{21}/5$.

57. Letting $t = \arccos \frac{5}{13}$, we want to find $\sin t$. Using $\cos t = \frac{5}{13}$ and $\sin^2 t + \cos^2 t = 1$, we see that

$$\sin^2 t + \left(\frac{5}{13}\right)^2 = 1 \qquad \text{or} \qquad \sin t = \sqrt{1 - \frac{25}{169}} = \sqrt{\frac{144}{169}} = \frac{12}{13}.$$

We used the positive square root because $\arccos \frac{5}{13}$ is in the first quadrant, so $\sin(\arccos \frac{5}{13})$ is positive. Thus, $\sin(\arccos \frac{5}{13}) = \frac{12}{13}$.

60. Letting $t = \tan^{-1} x$, we want to find $\sec t$. Using $\tan t = x$ and $1 + \tan^2 t = \sec^2 t$, we see that

$$1 + x^2 = \sec^2 t \qquad \text{or} \qquad \sec t = \sqrt{1 + x^2}.$$

Thus, $\sec(\tan^{-1} x) = \sqrt{1 + x^2}$. (We were able to use the positive square root because for $x \geq 0$, $0 \leq \tan^{-1} x < \pi/2$, and $t = \tan^{-1} x$ is in the first quadrant where $\sec t > 0$. On the other hand, for $x < 0$, $-\pi/2 < \tan^{-1} x < 0$, and $t = \tan^{-1} x$ is in the fourth quadrant where $\sec t > 0$.)

63. We note that the graph has amplitude $\frac{1}{2}$ and period 2π. We also note that it looks like the graph of $y = \cos x$. Since the graph is shifted up by 1 unit, it is $y = 1 + \frac{1}{2} \cos x$. Since $\cos x = \sin(\pi/2 - x)$, [see (9) in Section 4.2 of the text] we have $y = 1 + \frac{1}{2} \sin(x + \pi/2)$.

66. Let a denote the horizontal distance and c the straight-line distance as shown in the figure. Then, $\tan 43° = 20000/a$ and $\sin 43° = 20000/c$, so

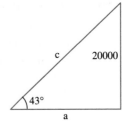

$$a = \frac{20000}{\tan 43°} = 21447.37 \text{ ft}$$

and

$$c = \frac{20000}{\sin 43°} = 29325.58 \text{ ft.}$$

69. **(a)** Refer to the figure in the text for this problem. Using the Law of Sines, we have

$$\frac{\sin \theta}{R} = \frac{\sin \rho}{H + R} \qquad \text{or} \qquad \sin \alpha = \frac{(36000 + 6370) \sin 6.5°}{6370} \approx 0.75,$$

where α is the related angle to ρ and ρ is the obtuse angle at point P. Then $\alpha = \sin^{-1} 0.75 \approx 48.85°$ and $\rho = 180° - \alpha = 180° - 48.85° = 131.15°$. Since the sum of the angles in a triangle is $180°$, we have

$$\phi = 180° - 6.5° - 131.15° = 42.35°.$$

(b) Let S be the distance from the satellite to the point P. Then, using the Law of Sines, we have

$$\frac{\sin \theta}{R} = \frac{\sin \phi}{S} \qquad \text{or} \qquad \sin \theta = \frac{R \sin \phi}{S}.$$

Using the Law of Cosines, we have

$$R^2 = S^2 + (H + R)^2 - 2S(H + R) \cos \theta$$

or

$$\cos \theta = \frac{S^2 + (H + R)^2 - R^2}{2S(H + R)}.$$

Dividing the above two results, we have

$$\tan \theta = \frac{\sin \theta}{\cos \theta} = \frac{R \sin \phi}{S} \bigg/ \frac{s^2 + (H + R)^2 - R^2}{2S(H + R)} = \frac{2(H + R)R \sin \phi}{S^2 + (H + R)^2 - R^2}.$$

To remove S^2 from this expression we again use the Law of Cosines:

$$S^2 = (H + R)^2 + R^2 - 2(H + R)R \cos \phi.$$

Substituting into the expression for $\tan \theta$, we have

$$\tan \theta = \frac{2(H + R)R \sin \phi}{(H + R)^2 + R^2 - 2(H + R)R \cos \phi + (H + R)^2 - R^2}$$

$$= \frac{2(H + R)R \sin \phi}{2(H + R)^2 - 2(H + R)R \cos \phi} = \frac{R \sin \phi}{H + R - R \cos \phi}$$

$$= \frac{R \sin \phi}{H + R(1 - \cos \phi)}.$$

72. In 30 minutes the airplane travels 200 miles and the car travels 30 miles. (See the figure to the right.) Using the right triangle with angle 6° and hypotenuse 200, we have

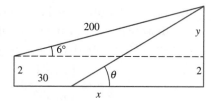

$$\cos 6° = \frac{x}{200}$$

or

$$x = 200 \cos 6° \approx 198.90.$$

The altitude of the airplane after 30 minutes is $y + 2$ where

$$\sin 6° = \frac{y}{200} \qquad \text{or} \qquad y = 200 \sin 6° = 20.91.$$

Thus, the altitude of the plane is $2 + 20.91 = 22.91$. The angle of elevation from the car to the airplane is θ, where

$$\tan \theta = \frac{22.91}{198.90} \approx 0.12 \qquad \text{so} \qquad \theta = \tan^{-1}(0.12) \approx 6.57°.$$

75. Refer to the figure in the text for this problem. The volume of the trough is $V = 20A$, where A is the area of the triangle. We need to express A as a function of θ. The base b of the triangle (on top in this case) is obtained from

$$\sin \theta = \frac{b/2}{4} \qquad \text{so} \qquad b = 2(4 \sin \theta) = 8 \sin \theta.$$

The height h of the triangle (the dashed line) is obtained from

$$\cos \theta = \frac{h}{4} \qquad \text{so} \qquad h = 4 \cos \theta.$$

The area of the triangle is

$$A = \frac{1}{2} bh = \frac{1}{2}(8 \sin \theta)(4 \cos \theta) = 16 \sin \theta \cos \theta = 8 \sin 2\theta,$$

and the volume is $V = 20A = 160 \sin 2\theta$.

78. Label the base of the triangle in the figure in the text x and the height of the triangle y. Then $A = \frac{1}{2} xy$ and we need to express x and y in terms of θ. Since $\cot \theta = x/y$,

$$x = y \cot \theta \qquad \text{and} \qquad A = \frac{1}{2}(y \cot \theta)y = \frac{1}{2} y^2 \cot \theta.$$

The perimeter of the triangle is $x + y + \sqrt{x^2 + y^2} = 2000$, so

$$y \cot \theta + y + \sqrt{(y \cot \theta)^2 + y^2} = y \cot \theta + y + y\sqrt{\cot^2 \theta + 1}$$

$$= y(\cot \theta + 1 + \sqrt{\csc^2 \theta}) = y(\cot \theta + 1 + \csc \theta) = 2000$$

and

$$y = \frac{2000}{1 + \cot \theta + \csc \theta}.$$

The area of the triangle is then

$$A(\theta) = \frac{1}{2} y^2 \cot \theta = \frac{1}{2} \cot \theta \cdot \left(\frac{2000}{1 + \cot \theta + \csc \theta} \right).$$

81. Let x and y be as shown in the figure to the right. The cross-section is a trapezoid with parallel sides having lengths 10 and $10 + 2x$. The area of the trapezoid is

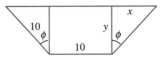

$$A = \frac{1}{2}[10 + (2x + 10)]y = \frac{1}{2}(20 + 2x)y = (10 + x)y,$$

so we need to express x and y in terms of ϕ. Using $\cos\phi = y/10$ and $\sin\phi = x/10$ we have $y = 10\cos\phi$ and $x = 10\sin\phi$, so

$$A = (10 + x)y = (10 + 10\sin\phi)(10\cos\phi)$$
$$= 100(1 + \sin\phi)\cos\phi = 100\cos\phi + 50\sin 2\phi.$$

Chapter 5

Exponential and Logarithmic Functions

5.1 $\boxed{\textbf{Exponential Functions}}$

3. Since $f(0) = -2^0 = -1$, the y-intercept is $(0, -1)$. The x-axis is a horizontal asymptote and the graph of $f(x) = -2^x$ is the graph of $y = 2^x$ reflected in the x-axis. The function is decreasing.

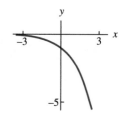

6. Since $f(0) = 2^{2-0} = 4$, the y-intercept is $(0, 4)$. The x-axis is a horizontal asymptote and the graph of $f(x) = 2^{2-x} = 2^{-(x-2)}$ is the graph of $y = 2^{-x}$ shifted 2 units to the right. The function is decreasing.

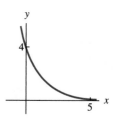

9. Since $f(0) = 3 - \left(\frac{1}{5}\right)^0 = 3 - 1 = 2$, the y-intercept is $(0, 2)$. The line $y = 3$ is a horizontal asymptote and the graph of $f(x) = 3 - \left(\frac{1}{5}\right)^x$ is the graph of $y = -\left(\frac{1}{5}\right)^x - 5^{-x}$ shifted up 3 units. The function is increasing.

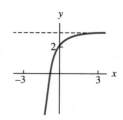

12. Since $f(0) = -3 - e^5$ the y-intercept is $(0, -3 - e^5)$. The line $y = -3$ is a horizontal asymptote and the graph of $f(x) = -3 - e^{x+5}$ is the graph of $y = -e^{-x}$ shifted down 3 units and to the left 5 units. The function is decreasing.

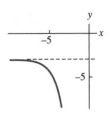

15. Letting $x = -1$ and $f(-1) = e^2$ we have $f(-1) = e^2 = b^{-1}$, so $b = e^{-2}$ and $f(x) = (e^{-2})^x = e^{-2x}$.

18. The range of $y = -2^{-x}$ is $(-\infty, 0)$, so the range of $f(x) = 4 - 2^{-x}$ is $(-\infty, 4)$.

21. Since $f(0) = 0e^0 + 10e^0 = 10$, the y-intercept is $(0, 10)$. To find any x-intercepts we solve $xe^x + 10e^x = (x + 10)e^x = 0$. Since $e^x > 0$ for all x, this implies $x + 10 = 0$ or $x = -10$. Thus, the only x-intercept is $(-10, 0)$.

24. Since

$$f(0) = \frac{2^0 - 6 + 2^{3-0}}{0 + 2} = \frac{1 - 6 + 8}{2} = \frac{3}{2},$$

the y-intercept is $(0, \frac{3}{2})$. To find any x-intercepts we solve

$$\frac{2^x - 6 + 2^{3-x}}{x + 2} = 0$$

$$2^x - 6 + 2^{3-x} = 0, \quad x \neq -2$$

$$2^{2x} - 6 \cdot 2^x + 2^3 = 0, \quad x \neq -2$$

$$(2^x)^2 - 6 \cdot 2^x + 8 = 0, \quad x \neq -2$$

$$(2^x - 4)(2^x - 2) = 0, \quad x \neq -2.$$

Then $2^x = 4$ so $x = 2$, and $2^x = 2$ so $x = 1$. The x-intercepts are $(2, 0)$ and $(1, 0)$.

27. Graphing $y = e^{x-2}$ and $y = 1$, we see that $e^{x-2} < 1$ for $x < 2$.

30. The graph of $f(x) = 3 - e^{-(x-1)^2}$ is the graph of $y = e^{-x^2}$ reflected through the x-axis, then shifted left 1 unit and up 3 units.

33. The graph of $f(x) = 1 - e^{x^2}$ is the graph of $y = e^{x^2}$ reflected through the x-axis and shifted up 1 unit.

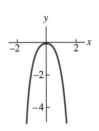

36. The graph of $f(x) = e^{(x+2)^2}$ is the graph of $y = e^{x^2}$ shifted 2 units to the left.

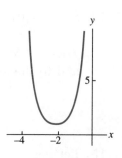

39. Writing

$$10^{-2x} = \frac{1}{1000} = \frac{1}{10^3} = 10^{-3},$$

we see that $-2x = -3$ so $x = \frac{3}{2}$.

42. Writing

$$3^x = 27\left(\frac{1}{3}\right)^x = 3^3\left(\frac{1}{3x}\right) = 3^{3-x},$$

we see that $x = 3 - x$, $2x = 3$, and $x = \frac{3}{2}$.

45. Writing

$$3^x = 27^{x^2} = (3^3)^{x^2} = 3^{3x^2},$$

we see that $x = 3x^2$ or $x(3x - 1) = 0$. Thus, $x = 0$ or $x = \frac{1}{3}$.

48. We write the equation as

$$(8^2)^x - 10(8^x) + 16 = (8^x)^2 - 10(8^x) + 16 = (8^x - 2)(8^x - 8) = 0.$$

Then

$$8^x = 2, \quad (2^3)^x = 2, \quad 2^{3x} = 2, \quad \text{or} \quad 3x = 1$$

and $x = \frac{1}{3}$. Also, $8^x = 8$ or $x = 1$. The solutions are $x = \frac{1}{3}$ and $x = 1$.

51. The x-intercept is the solution of $e^{x+4} - e = 0$ or $e^{x+4} = e^1$. Thus, $x + 4 = 1$ and $x = -3$, so the x-intercept is $(-3, 0)$.

54.

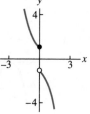

5.2 Logarithmic Functions

3. $10^4 = 10,000$ is equivalent to $\log_{10} 10,000 = 4$.

6. $(a+b)^2 = a^2 + 2ab + b^2$ is equivalent to $\log_{a+b}(a^2 + 2ab + b^2) = 2$.

9. $\log_{\sqrt{3}} 81 = 8$ is equivalent to $(\sqrt{3})^8 = 81$.

12. $\log_b b^2 = 2$ is equivalent to $b^2 = b^2$.

15. $\log_2(2^2 + 2^2) = \log_2(4 + 4) = \log_2 8 = \log_2 2^3 = 3$

18. $\ln(e^4 e^9) = \ln e^{4+9} = \ln e^{13} = 13$

21. $e^{-\ln 7} = e^{\ln 7^{-1}} = 7^{-1} = \frac{1}{7}$

24. We solve $\frac{1}{3} = \ln_b 4$ or $b^{1/3} = 4$. Cubing both sides, we have $b = 4^3 = 64$

27. The domain of $\log_2(x)$ is determined by $x > 0$, so the domain of $\log_2(-x)$ is determined by $-x > 0$ or $x < 0$. That is, the domain of $\log_2(-x)$ is $(-\infty, 0)$. The x-intercept is the solution of $\log_2(-x) = 0$. This is equivalent to $2^0 = -x$ or $x = -2^0 = -1$, so the x-intercept is $(-1, 0)$. The vertical asymptote is $-x = 0$ or $x = 0$, which is the y-axis.

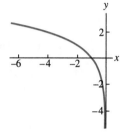

30. The domain of $\log_4 x$ is determined by $x > 0$, so the domain of $\log_4(x - 4)$ is determined by $x - 4 > 0$ or $x > 4$. Thus, the domain of $1 - 2\log_4(x - 4)$ is $(4, \infty)$. The x-intercept is the solution of

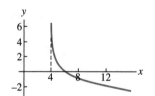

$$1 - 2\log_4(x - 4) = 0$$
$$2\log_4(x - 4) = 1$$
$$\log_4(x - 4) = \frac{1}{2}$$
$$4^{1/2} = x - 4$$
$$x = 4 + 2 = 6,$$

so the x–intercept is $(6, 0)$. The vertical asymptote is $x - 4 = 0$ or $x = 4$.

33. From the graph, we see that $\ln(x + 1) < 0$ for $-1 < x < 0$.

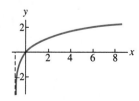

36. The graph of $y = \ln|x - 2|$, shown dashed, is the graph of $y = \ln|x|$, shown in blue, shifted 2 units to the right. The x-intercepts occur where $\ln|x - 2| = 0$ or $|x - 2| = 1$. They are $(1, 0)$ and $(3, 0)$. The vertical asymptote is $y = 2$.

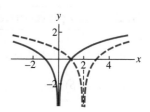

39. The domain of $\ln(2x - 1)$ is determined by $2x - 3 > 0$ or $x > \frac{3}{2}$. Thus, the domain is $(\frac{3}{2}, \infty)$.

42. The domain of $\ln(x^2 - 2x)$ is determined by $x^2 - 2x > 0$. This is equivalent to $x(x - 2) > 0$. From the sign chart below

x	$-$	0	$+$	$+$	$+$
$x - 2$	$-$	$-$	$-$	0	$+$
$x(x-2)$	$+$	0	$-$	0	$+$

we see that the domain is $(-\infty, 0) \cup (2, \infty)$.

45. Using (ii) of the Laws of Logarithms in the text (reading the law from right to left), we have

$$\ln(x^4 - 4) - \ln(x^2 + 2) = \ln \frac{x^4 - 4}{x^2 + 2} = \ln \frac{(x^2 - 2)(x^2 + 2)}{x^2 + 2} = \ln(x^2 - 2).$$

48. Using (i), (ii), and (iii) of the Laws of Logarithms in the text (reading the laws from right to left), we have

$$5\ln 2 + 2\ln 3 - 3\ln 4 = \ln 2^5 + \ln 3^2 - \ln 4^3 = \ln \frac{2^5 \cdot 3^2}{4^3} = \ln \frac{32 \cdot 9}{64} = \ln \frac{9}{2}.$$

51. Using the Laws of Logarithms in the text, we have

$$\ln y = \ln \frac{(x^3 - 3)^5(x^4 + 3x^2 + 1)^8}{\sqrt{x}(7x + 5)^9}$$
$$= \ln(x^3 - 3)^5 + \ln(x^4 + 3x^2 + 1)^8 - \ln x^{1/2} - \ln(7x + 5)^9$$
$$= 5\ln(x^3 - 3) + 8\ln(x^4 + 3x^2 + 1) - \frac{1}{2}\ln x - 9\ln(7x + 5).$$

54. We use the Laws of Logarithms and the fact that $\ln x$ is one-to-one:

$$\ln 3 + \ln(2x - 1) = \ln 4 + \ln(x + 1)$$
$$\ln[3(2x - 1)] = \ln[4(x + 1)]$$
$$3(2x - 1) = 4(x + 1)$$
$$6x - 3 = 4x + 4$$
$$2x = 7$$
$$x = \frac{7}{2}.$$

Since solutions of equations involving logarithms may be extraneous, it is necessary to check your answer. (See Part 18 in **Topics in Algebra**.) In this case, substituting $x = \frac{7}{2}$ into the equation, we have

$$\ln 3 + \ln\left(2 \cdot \frac{7}{2} - 1\right) = \ln 3 + \ln 6 = \ln 3 + \ln(2 \cdot 3) = \ln 3 + \ln 2 + \ln 3 = \ln 2 + 2\ln 3$$

and

$$\ln 4 + \ln\left(\frac{7}{2} + 1\right) = \ln 2^2 + \ln\frac{9}{2} = 2\ln 2 + \ln 9 - \ln 2 = \ln 2 + 2\ln 3.$$

Thus, the solution checks.

57. We use the Laws of Logarithms and the fact that $\log_2 x$ is one-to-one:

$$\log_2(x - 3) - \log_2(2x + 1) = -\log_2 4$$
$$\log_2\frac{x - 3}{2x - 1} = \log_2 4^{-1}$$
$$\frac{x - 3}{2x + 1} = \frac{1}{4}$$
$$4x - 12 = 2x + 1$$
$$2x = 13$$
$$x = \frac{13}{2}.$$

Checking this result, we have

$$\log_2\left(\frac{13}{2} - 3\right) - \log_2\left(2 \cdot \frac{13}{2} + 1\right) = \log_2\frac{7}{2} - \log_2 14 = \log_2\frac{7/2}{14} = \log_2\frac{7}{28}$$
$$= \log_2\frac{1}{4} = \log_2 4^{-1} = -\log_2 4.$$

Thus, the solution checks.

60. Writing 2^{2x} as $(2^x)^2$, we have

$$2^{2x} - 12(2^x) + 35 = (2x)^2 - 12(2^x) + 35 = (2^x - 7)(2^x - 5) = 0.$$

Thus, $2^x = 7$ and $2^x = 5$. Taking the logarithm base of 2 of both equations, we have $x = \log_2 7$ and $x = \log_2 5$.

63. Rewriting the equation in exponential form, we have

$$\log_{10}\frac{1}{x} = 2$$
$$\frac{1}{x} = 10^2 = 100$$
$$x = \frac{1}{100}.$$

66. Rewriting the equation in exponential form, we have

$$\log_5 |1 - x| = 1$$
$$|1 - x| = 5^1 = 5.$$

Thus,

$$1 - x = 5 \qquad \text{or} \qquad 1 - x = -5$$
$$x = -4 \qquad \text{or} \qquad x = 6.$$

69. Using properties of logarithms, we have

$$\log_3 81^x - \log_3 3^{2x} = 3$$
$$\log_3 (3^4)^x - \log_3 3^{2x} = 3$$
$$\log_3 3^{4x} - \log_3 3^{2x} = 3$$
$$4x - 2x = 3$$
$$2x = 3$$
$$x = \frac{3}{2}.$$

This answer checks.

72. We want to solve $\left(\frac{1}{2}\right)^x = 7$. This is equivalent to

$$\frac{1}{2^x} = 7$$
$$2^x = \frac{1}{7}$$
$$\ln 2^x = \ln \frac{1}{7} = \ln 7^{-1} = -\ln 7$$
$$x \ln 2 = -\ln 7$$
$$x = -\frac{\ln 7}{\ln 2}.$$

75. Taking the natural logarithm of both sides, we have

$$\ln 5^x = \ln(2e^{x+1}) = \ln 2 + \ln e^{x+1} = \ln 2 + x + 1$$
$$x \ln 5 = \ln 2 + x + 1$$
$$x \ln 5 - x = 1 + \ln 2$$
$$x(\ln 5 - 1) = 1 + \ln 2$$
$$x = \frac{1 + \ln 2}{\ln 5 - 1}.$$

5.3   Exponential and Logarithmic Models

3. The initial population is 1500, so an increase of 25% is $0.25(1500) = 375$. Thus, $P(10) = 1500e^{10k} = 1875$. Solving for k gives

$$e^{10k} = \frac{1875}{1500}, \qquad 10k = \ln \frac{1875}{1500} \approx 0.2231, \qquad \text{or} \qquad k = 0.0223.$$

Thus, $P(t) = 1500e^{0.0223t}$ and $P(20) = 1500e^{0.0223(2)} \approx 2344$.

6. **(a)** We are given $P(2) = 200$ and $P(5) = 400$. That is,

$$200 = P_0 e^{2k} \quad \text{and} \quad 400 = P_0 e^{5k}.$$

Dividing, we have $e^{5k}/e^{2k} = 400/200 = 2$ or $e^{3k} = 2$. Thus, $3k = \ln 2$ and $k = \frac{1}{3}\ln 2 \approx 0.2310$. Now, solving $P(2) = P_0 E^{0.2310(2)} = 200$ for P_0 we find $P_0 = 200e^{-0.4621} \approx 126$. Thus, the model is $P(t) = 126e^{0.2310t}$.

(b) In ten days $P(10) \approx 1270$.

(c) Solving $P(t) = 126e^{0.2310t} = 5000$ for t, we have

$$e^{0.2310t} = \frac{5000}{126}$$

$$0.2310t = \ln \frac{5000}{126}$$

$$t = \frac{1}{0.2310} \ln \frac{5000}{126} \approx 15.9 \text{ days.}$$

9. We are given $A_0 = 200$, so $A(t) = 200e^{kt}$. Since 3% of 200 is 6, $A(6) = 200e^{6k} = 194$ and

$$e^{6k} = \frac{194}{200}, \qquad 6k = \ln \frac{194}{200}, \qquad \text{or} \qquad k = \frac{1}{6}\ln \frac{199}{200} \approx -0.0051.$$

Thus, $A(t) = 200e^{-0.0051t}$ and $A(24) \approx 177.1$ mg.

12. We are given $A_0 = 400$, so $A(t) = 400e^{kt}$. After 8 hours, one-half, or 200 grams, remains. Solving $A(8) = 400e^{8k} = 200$ for k we have

$$e^{8k} = \frac{200}{400} = \frac{1}{2}, \qquad 8k = \ln \frac{1}{2}, \qquad \text{or} \qquad k = \frac{1}{8}\ln \frac{1}{2} \approx -0.0866.$$

Thus, $A(t) = 400e^{-0.0866t}$ so $A(17) \approx 91.7$, $A(23) \approx 54.5$, and $A(33) \approx 22.9$.

15. The initial amount is not given so we represent it using A_0 and express the remaining amounts in terms of A_0. Since the half-life is 140 days, $A(140) = A_0 e^{140k} = \frac{1}{2}A_0$, which we solve for k:

$$e^{140k} = \frac{1}{2}, \qquad 140k = \ln \frac{1}{2}, \qquad \text{or} \qquad k = \frac{1}{140}\ln \frac{1}{2} \approx -0.0050.$$

Thus, $A(t) = A_0 e^{-0.0050t}$ and

$$A(80) = A_0 e^{-0.0050(80)} \approx 0.6730A_0$$
$$A(300) = A_0 e^{-0.0050(300)} \approx 0.2264A_0.$$

18. From Example 4 in the text, we see that the amount of C-14 remaining at t years is given by $A(t) = A_0 e^{-0.00012097t}$, where A_0 is the initial amount of C-14 in the organism. We want to solve this equation for t when $A(t) = 0.10A_0$, and again when $A(t) = 0.05A_0$. (When 90% is lost, it means that 10% is remaining.) This gives

$$A_0 e^{-0.00012097t} = 0.10A_0$$
$$-0.00012097t = \ln 0.1$$
$$t = -\frac{1}{0.00012097} \ln 0.1 \approx 19{,}034$$

and

$$A_0 e^{-0.00012097t} = 0.05A_0$$
$$-0.00012097t = \ln 0.05$$
$$t = -\frac{1}{0.00012097} \ln 0.05 \approx 24{,}764,$$

respectively, Thus, the bone is between 19,034 and 24,764 years old.

21. We identify $T_0 = 400$, $T_m = 80$, and $T(3) = 275$, where $T(t) = T_m - (T_0 - T_m)e^{kt}$. Using this information, we solve for k:

$$T(3) = 80 + (400 - 80)e^{3k} = 275$$
$$320e^{3k} = 195$$
$$e^{3k} = \frac{195}{320}$$
$$3k = \ln \frac{195}{320}$$
$$k = \frac{1}{3} \ln \frac{195}{320} \approx -0.1651.$$

Thus, $T(t) = 80 + 320e^{-0.1651t}$.

(a) $T(5) \approx 220.2°\text{F}$

(b) We solve $T(t) = 150$ for t:

$$80 + 320e^{-0.1651t} = 150$$
$$320e^{-0.1651t} = 70$$
$$e^{-0.1651t} = \frac{7}{32}$$
$$-0.1651t = \ln \frac{7}{32}$$
$$t = -\frac{1}{0.1651} \ln \frac{7}{32} \approx 9.2 \text{ min.}$$

(c) After a very long time, $e^{-0.1651t}$ approaches 0, so T approaches $80°\text{F}$.

24. We identify $T_m = 5$, $T(1) = 59$, and $T(5) = 32$. Then

$$5 + (T_0 - 5)e^k = 59 \qquad \text{and} \qquad 5 + (T_0 - 5)e^{5k} = 32$$

or

$$(T_0 - 5)e^k = 54 \qquad \text{and} \qquad (T_0 - 5)e^{5k} = 27.$$

Dividing, we have $e^{4k} = \frac{1}{2}$, so $e^k = \left(\frac{1}{2}\right)^{1/4}$. Then

$$(T_0 - 5)\left(\frac{1}{2}\right)^{1/4} = 54 \qquad \text{or} \qquad T_0 = 54\left(\frac{1}{2}\right)^{-1/4} + 5 \approx 69.2°\text{F}.$$

Thus, the temperature inside the house is $69.2°\text{F}$.

27. Using $S(t) = Pe^{rt}$ we identify $P = 0.01$ and $r = 0.01$. Then $S(2000) = 0.01e^{0.01(2000)} \approx$ \$4,851,651.95. If $r = 0.02$, then

$$S(2000) = 0.01e^{0.02(2000)} \approx 2{,}353{,}852{,}668{,}370{,}199.85 \approx 2.35 \times 10^{15}.$$

30. Using $P(t) = Se^{-rt}$ we identify $S = 100{,}000$, $r = 0.03$, and $t = 30$. then

$$P(30) = 100{,}000e^{-0.03(30)} \approx \$40{,}656.97.$$

33. From

$$I = \frac{E}{R}\left(1 - e^{-(R/L)t}\right)$$

we have

$$\frac{IR}{E} = 1 - e^{-(R/L)t}$$

$$e^{-(R/L)t} = 1 - \frac{IR}{E}$$

$$-\frac{R}{L}t = \ln\left(1 - \frac{IR}{E}\right)$$

$$t = -\frac{L}{R}\ln\left(1 - \frac{IR}{E}\right).$$

36. The magnitude of an earthquake is defined by $M = \log_{10}(A/A_0)$, where A and A_0 are amplitudes of seismic waves. Using this formula, we have

$$9.3 = \log_{10}\left(\frac{A}{A_0}\right)_{\text{Sumatra}} \qquad \text{and} \qquad 8.9 = \log_{10}\left(\frac{A}{A_0}\right)_{\text{Alaska}}.$$

Then

$$\left(\frac{A}{A_0}\right)_{\text{Sumatra}} = 10^{9.3} \qquad \text{and} \qquad \left(\frac{A}{A_0}\right)_{\text{Alaska}} = 10^{8.9}.$$

Since $9.3 = 8.9 + 0.4$, it follows from the laws of exponents that

$$\left(\frac{A}{A_0}\right)_{\text{Sumatra}} = 10^{9.3} = 10^{0.4}10^{8.9} = 10^{0.4}\left(\frac{A}{A_0}\right)_{\text{Alaska}} \approx 2.51\left(\frac{A}{A_0}\right)_{\text{Alaska}}.$$

Thus, the Sumatra earthquake was about $2\frac{1}{2}$ times as intense as the Alaska earthquake.

39. Since $pH = -\log_{10}[H^+]$ where $[H^+]$ is the hydrogen-ion concentration, we have

$$pH(10^{-6}) = -\log_{10}10^{-6} = 6.$$

42. Since $pH = -\log_{10}[H^+]$ where $[H^+]$ is the hydrogen-ion concentration, we have

$$pH(5.1 \times 10^{-5}) = -\log_{10}(5.1 \times 10^{-5}) = -\log_{10}5.1 - \log_{10}10^{-5}$$
$$= 5 - \log_{10}5.1 \approx 5 - 0.71 = 4.29.$$

45. We need to solve $6.6 = -\log_{10}[H^+]$:

$$-6.6 = \log_{10}[H^+]$$
$$10^{-6.6} = [H^+]$$
$$[H^+] \approx 2.5 \times 10^{-7},$$

48. From the formula for pH we have

$$2.3 = -\log_{10}[H^+]_{\text{lemon juice}} \qquad \text{and} \qquad 3.3 = -\log_{10}[H^+]_{\text{vinegar}}.$$

Then

$$[H^+]_{\text{lemon juice}} = 10^{-2.3} \qquad \text{and} \qquad [H^+]_{\text{vinegar}} = 10^{-3.3}.$$

Since $-2.3 = 1 - 3.3$, it follows from the laws of exponents that

$$[H^+]_{\text{lemon juice}} = 10^{-2.3} = 10^1 10^{-3.3} = 10[H^+]_{\text{vinegar}}.$$

Thus, lemon juice is 10 times more acidic than vinegar.

51. We need to solve $M = \frac{2}{3}[\log_{10} E - 11.8]$ for E given that $M = 9.3$:

$$\frac{2}{3}[\log_{10} E - 11.8] = 9.3$$
$$\log_{10} E - 11.8 = \frac{3}{2}(9.3) = 13.95$$
$$\log_{10} E = 25.7$$
$$E = 10^{25.75} \approx 5.62 \times 10^{25} \text{ ergs.}$$

54. From (10) in the text we have

$$b_1 = 10\log_{10}\frac{I_1}{I_0} \qquad \text{and} \qquad b_2 = 10\log_{10}\frac{I_2}{I_0}.$$

Then

$$b_2 - b_1 = 10\log_{10}\frac{I_2}{I_0} - 10\log_{10}\frac{I_1}{I_0} = 10\log_{10}\frac{I_2/I_0}{I_1/I_0} = 10\log_{10}\frac{I_2}{I_1}.$$

From (11) in the text, we have

$$I_1 = \frac{k}{d_1^2} \qquad \text{and} \qquad I_2 = \frac{k}{d_2^2},$$

so

$$b_2 - b_1 = 10 \log_{10} \frac{I_2}{I_1} = 10 \log_{10} \frac{k/d_2^2}{k/d_1^2} = 10 \log_{10} \left(\frac{d_1}{d_2} \right)^2 = 20 \log_{10} \frac{d_1}{d_2}.$$

Thus,

$$b_2 = b_1 + 2 \log_{10} \frac{d_1}{d_2}.$$

57. **(a)** When d is the pupil diameter and $B = 255$ we have

$$\log_{10} d = 0.8558 - 0.000401(8.1 + \log_{10} 255)^3 \approx 0.3907,$$

so

$$d = 10^{0.3907} \approx 2.46 \text{ mm}.$$

(b) From $\log_{10} d = -0.10446$ when $B = 190{,}000$ mL we have $d = 10^{-0.10446} \approx 0.79$ mm, and from $\log_{10} d = -0.728144$ when $B = 51{,}000{,}000$ mL we have $d = 10^{-0.728144} \approx 0.19$ mm.

(c) We want to solve

$$\log_{10} 7 = 0.8558 - 0.000401(8.1 + \log_{10} B)^3$$

for B:

$$0.8451 = 0.8558 - 0.000401(8.1 + \log_{10} B)^3$$
$$0.000401(8.1 + \log_{10} B)^3 = 0.0107$$
$$(8.1 + \log_{10} B)^3 = \frac{0.0107}{0.000401} \approx 26.6882$$
$$8.1 + \log_{10} B = 26.6882^{1/3} \approx 2.9884$$
$$\log_{10} B = 2.9884 - 8.1 = -5.1116$$
$$B = 10^{-5.1116} \approx 7.73 \times 10^{-6} \text{ mL}.$$

5.4 | Calculus Preview—The Number e |

3. From Problem 2, we have that the derivative of $f(x) = \log_b x$ is $f'(x) = \frac{1}{x} \log_b e$. Thus, when $f(x) = \log_{10} x$, we have $f'(x) = \frac{1}{x} \log_{10} e$. Using (11) from Section 5.2 in the text, we can also write

$$f'(x) = \frac{1}{x} \log_{10} e = \frac{1}{x} \frac{\ln e}{\ln 10} = \frac{1}{x \ln 10}.$$

6. We first write $f(x) = \log_{10} 6x = \log_{10} 6 + \log_{10} x$. We then compute the difference quotient:

$$\frac{f(x+h) - f(x)}{h} = \frac{\log_{10} 6 + \log_{10}(x+h) - [\log_{10} 6 + \log_{10} x]}{h}$$
$$= \frac{\log_{10}(x+h) - \log_{10} x}{h}.$$

Thus,

$$\lim_{h \to 0} \frac{f(x+h) - f(x)}{h} = \lim_{h \to 0} \frac{\log_{10}(x+h) - \log_{10} x}{h},$$

which we recognize as the derivative of $\log_{10} x$. By Problem 3, this is $1/x \ln 10$, so the derivative of $f(x) = \log_{10} 6x$ is $f'(x) = 1/x \ln x$.

9. Using $\cosh x = \frac{1}{2}(e^x + e^{-x})$ and $\sinh x = \frac{1}{2}(e^x - e^{-x})$, we have

$$\cosh^2 x - \sinh^2 x = \frac{1}{4}(e^x + e^{-x})^2 - \frac{1}{4}(e^x - e^{-x})^2$$

$$= \frac{1}{4}(e^{2x} + 2 + e^{-2x}) - \frac{1}{4}(e^{2x} - 2 + e^{-2x})$$

$$= \frac{2}{4} + \frac{2}{4} = 1.$$

12. Using $\sinh x = \frac{1}{2}(e^x - e^{-x})$, we have

$$\sinh(-x) = \frac{1}{2}\left(e^{-x} - e^{-(-x)}\right) = -\frac{1}{2}(e^x - e^{-x}) = -\sinh x.$$

15. Given that $\cosh x = \frac{3}{2}$ and using $\cosh^2 x - \sinh^2 x = 1$, we have

$$\sinh^2 x = \cosh^2 x - 1 = \left(\frac{3}{2}\right)^2 - 1 = \frac{9}{4} - 1 = \frac{5}{4}.$$

Thus, $\sinh x = \pm\sqrt{5}/2$.

18. Letting $y = \cosh x = \frac{1}{2}(e^x - e^{-x})$, we have

$$e^x + e^{-x} = 2y \qquad \text{or} \qquad e^x - 2y + e^{-x} = 0.$$

Multiplying both sides by e^x, we find

$$(e^x)^2 - 2ye^x + 1 = 0,$$

which is quadratic in e^x, so we use the quadratic formula:

$$e^x = \frac{2y \pm \sqrt{4y^2 - 4}}{2} = y \pm \sqrt{y^2 - 1}.$$

Since $x > 0$, $e^x > 1$, and we must use $e^x = y + \sqrt{y^2 - 1}$. $\Big($Because $y > 1$,

$$2y > 2$$
$$-2y < -2$$
$$y^2 - 2y + 1 < y^2 - 2 + 1 = y^2 - 1$$
$$(y - 1)^2 < y^2 - 1$$
$$y - 1 < \sqrt{y^2 - 1}$$
$$y - \sqrt{y^2 - 1} < 1. \Big)$$

Taking the natural logarithm of both sides of $e^x = y + \sqrt{y^2 - 1}$, we have $x = \ln\left(y + \sqrt{y^2 - 1}\right)$. Thus, the inverse of $y = \cosh x$, $x \ge 0$, is $y = \ln\left(x + \sqrt{x^2 - 1}\right)$.

Chapter 5 Review Exercises

3. Since $y = \ln x$ has x-intercept at 1 and vertical asymptote $x = -4$, we use the fact that the graph of $y = \ln(x + 4)$ is the graph of $y = \ln x$ shifted 4 units to the left to see that $y = \ln(x + 4)$ has x-intercept $(-3, 0)$ and vertical asymptote $x = -4$.

6. $6 \ln e + 3 \ln \frac{1}{e} = 6(1) + 3 \ln e^{-1} = 6 - 3 \ln e = 6 - 3 = 3$

9. $\log_4(4 \cdot 4^2 \cdot 4^3) = \log_4 4^{1+2+3} = 1 + 2 + 3 = 6$

12. If $\log_b 6 = \frac{1}{2}$, then $b^{1/2} = 6$ and $b = 36$.

15. If $-1 + \ln(x - 3) = 0$, then $\ln(x - 3) = 1$, $x - 3 = e$, and $x = 3 + e$.

18. If $3^x = 5$, then $3^{-2x} = (3^x)^{-2} = 5^{-2} = \frac{1}{5^2} = \frac{1}{25}$.

21. We let $x = 2$ and $y = 9$ and solve for C:

$$9 = e^{2-2} + c = 1 + c$$
$$c = 8.$$

24. True; since $\log_b b = 1$.

27. True; since $4^{x/2} = (4^{1/2})^x = 2^x$.

30. True; since $2^{3+3x} = 2^{3(1+x)} = (2^3)^{1+x} = 8^{1+x}$.

33. False; since $\ln(\ln e) = \ln 1 = 0$.

36. True; since $\log_6(36)^{-1} = \log_6(6^2)^{-1} = \log_6 6^{-2} = -2$.

39. $\log_9 27 = 1.5$ is equivalent to $9^{1.5} = 27$.

42. Since $3^{2x} = 81 = 3^4$, $2x = 4$ and $x = 2$.

45. Taking the logarithm base 2 of both sides:

$$\log_2 2^{1-x} = \log_2 7, \qquad 1 - x = \log_2 7, \qquad \text{and} \qquad x = 1 - \log_2 7.$$

Alternatively, taking the natural logarithm of both sides:

$$\ln 2^{1-x} = \ln 7, \qquad (1 - x) \ln 2 = \ln 7, \qquad \text{and} \qquad x = 1 - \frac{\ln 7}{\ln 2}.$$

48. Dividing both sides of the equation by e^x:

$$3 = 4 \frac{e^{-3x}}{e^x} = 4e^{-4x}$$
$$e^{-4x} = \frac{3}{4}$$
$$-4x = \ln \frac{3}{4}$$
$$x = -\frac{1}{4} \ln \frac{3}{4} \approx 0.0719.$$

51. When $b > 2$, the graph of $f(x) = b^x$ is C. When $1 < b < 2$, the graph of b^x is D. When $\frac{1}{2} < b < 1$, the graph of b^x is A. When $0 < b < \frac{1}{2}$, the graph of b^x is B. Thus, (i) is C, (ii) is D, (iii) is A, and (iv) is B.

54. Since $\ln 1 = 0$ and $\ln e^2 = 2$, the line passes through $(1, 0)$ and $(e^2, 2)$. Its slope is

$$m = \frac{2 - 0}{e^2 - 1} = \frac{2}{e^2 - 1}.$$

57. This looks like the graph of $y = \ln x$ revolved around the x-axis (giving $-\ln x$) and shifted left 2 units (giving $-\ln(x + 2)$). But then the graph should pass through $(-1, 0)$. Instead, it appears to pass through $(-1, -2)$. Thus, this is the graph of (iv) $y = -2 - \ln(x + 2)$.

60. This looks like the graph of $y = \ln(-x)$ shifted right by 2 units. This would be $y = \ln[-(x - 2)] = \ln(-x + 2)$. But then the graph should pass through $(1, 0)$. Instead, it appears to pass through $(1, 2)$. Thus, this is the graph of (vi) $y = 2 + \ln(-x + 2)$.

63. Since $(0, 5)$ is on the graph, $5 = Ae^{k \cdot 0} = A$ and the function is $f(x) = 5e^{kx}$. Since the graph passes through $(6, 1)$, $1 = 5e^{6k}$ and $k = \frac{1}{6} \ln \frac{1}{5} = -\frac{1}{6} \ln 5$. Thus, $f(x) = 5e^{-(\frac{1}{6} \ln 5)x} = 5e^{-0.2682x}$.

66. The graph of $y = \log_3 x$ has a vertical asymptote at $x = 0$, so the graph of $y = a + \log_3(x - 2)$ has a vertical asymptote at $x = 2$. The graph passes through $(11, 10)$, so

$$10 = a + \log_3(11 - 2) = a + \log_3 9 = a + \log_3 3^2 = a + 2$$

and $a = 8$. Thus, $f(x) = 8 + \log_3(x - 2)$.

69. If A_0 is the initial amount and $\frac{1}{2}A_0$ remains after 12.5 years, then $\frac{1}{2}(\frac{1}{2}A_0) = \frac{1}{4}A_0$ remains after $25 = 2(12.5)$ years and $\frac{1}{4}(\frac{1}{4}A_0) = \frac{1}{16}A_0 = 0.0625A_0$ remains after $50 = 2(25)$ years.

72. We are given that $I(1) = 0.30I_0 = I_0 e^{k \cdot 1} = I_0 e^k$. Thus, $0.3 = e^k$ and

$$I(3) = I_0 e^{k \cdot 3} = I_0(e^k)^3 = I_0(0.3)^3 = 0.027I_0.$$

That is, the intensity 3 meters below the surface is 2.7% of I_0.

Chapter 6

Conic Sections

6.1 | The Parabola |

3. The form of the parabola is $y^2 = 4cx$, $c < 0$, so the parabola opens to the left, and $4c = -\frac{4}{3}$ so $c = -\frac{1}{3}$. The vertex is at the origin, the focus is at $F(c, 0)$ or $(-\frac{1}{3}, 0)$, the directrix is $x = -c = \frac{1}{3}$, and the axis is $y = 0$. The ends of the focal chord are $(-\frac{1}{3}, \pm\frac{2}{3})$. The graph of the parabola is shown, along with its focus F and directrix.

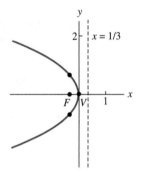

6. The form of the parabola is $x^2 = 4cy$, $c > 0$, so the parabola opens up, and $4c = \frac{1}{10}$ so $c = \frac{1}{40}$. The vertex is at the origin, the focus is at $F(0, c)$ or $(0, \frac{1}{40})$, the directrix is $x = -c = -\frac{1}{40}$, and the axis is $x = 0$. The ends of the focal chord are $(\pm\frac{1}{20}, \frac{1}{40})$. The graph of the parabola is shown, along with its focus F and directrix.

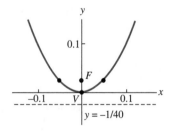

9. The form of the parabola is $(y - k)^2 = 4c(x - h)$, $c > 0$, so
the parabola opens to the right, and $4c = 16$ so $c = 4$. The
vertex is at $(0, 1)$, the focus is at $F(h + c, k)$ or $(4, 1)$, the
directrix is $x = -c = -4$, and the axis is $y = 1$. The ends
of the focal chord are $(4, \pm 8)$. The graph of the parabola
is shown, along with its focus F and directrix.

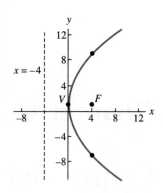

12. The form of the parabola is $(x - h)^2 = 4c(y - k)$,
$c < 0$, so the parabola opens down, and $4c = -1$
so $c = -\frac{1}{4}$. The vertex is at $(2, 0)$, the focus is
at $F(2, c)$ or $(2, -\frac{1}{4})$, the directrix is $y = -c = $
$\frac{1}{4}$, and the axis is $x = 2$. The ends of the focal
chord are $(2 \pm \frac{1}{2}, -\frac{1}{4})$. The graph of the parabola
is shown, along with its focus F and directrix.

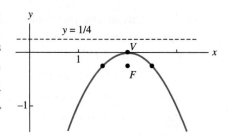

15. We rewrite the equation in standard form (3) in the
text:

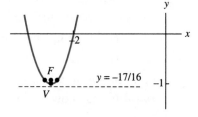

$$x^2 + 5x = \frac{1}{4}y - 6$$

$$x^2 + 5x + \frac{25}{4} = \frac{1}{4}y - 6 + \frac{25}{4}$$

$$\left(x + \frac{5}{2}\right)^2 = \frac{1}{4}y + \frac{1}{4} = 4\left(\frac{1}{16}\right)(y + 1).$$

This is a parabola that opens up, with $c = \frac{1}{16}$. The vertex is at $(-\frac{5}{2}, -1)$, the focus
is at $F(-\frac{5}{2}, -1 + \frac{1}{16})$ or $(-\frac{5}{2}, -\frac{15}{16})$, the directrix is $y = -1 - \frac{1}{16} = -\frac{17}{16}$, and the axis
is $x = -\frac{5}{2}$. The ends of the focal chord are $(-\frac{5}{2} \pm \frac{1}{8}, -\frac{15}{16})$. The graph of the parabola
is shown, along with its focus F and directrix.

18. We rewrite the equation in standard form (4) in the text:

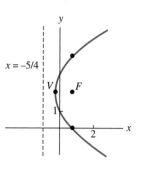

$$y^2 - 4y \quad\; = 4x - 3$$
$$y^2 - 4y + 4 = 4x - 3 + 4$$
$$(y-2)^2 = 4x + 1 = 4\left(x + \frac{1}{4}\right).$$

This is a parabola that opens to the right, with $c = 1$. The vertex is at $\left(-\frac{1}{4}, 2\right)$, the focus is at $F\left(-\frac{1}{4}+1, 2\right)$ or $\left(\frac{3}{4}, 2\right)$, the directrix is $x = -\frac{1}{4} - 1 = -\frac{5}{4}$, and the axis is $y = 2$. The ends of the focal chord are $\left(\frac{3}{4}, 2 \pm 2\right)$. The graph of the parabola is shown, along with its focus F and directrix.

21. We rewrite the equation in standard form (3) in the text:

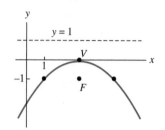

$$2x^2 + 12x \quad\; = 8y + 18$$
$$x^2 - 6x \quad\; = -4y - 9$$
$$x^2 - 6x + 9 = -4y - 9 + 9$$
$$(x-3)^2 = -4y.$$

This is a parabola that opens down, with $c = -1$. The vertex is at $(3, 0)$, the focus is at $F(3, -1)$, the directrix is $y = 1$, and the axis is $x = 3$. The ends of the focal chord are $(3 \pm 2, -1)$. The graph of the parabola is shown, along with its focus F and directrix.

24. We rewrite the equation in standard form (3) in the text:

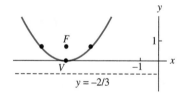

$$3x^2 + 30x \quad\; = 8y - 75$$
$$x^2 + 10x \quad\; = \frac{8}{3}y - 25$$
$$x^2 + 10x + 25 = \frac{8}{3}y - 25 + 25$$
$$(x+5)^2 = 4\left(\frac{2}{3}\right)y.$$

This is a parabola that opens up, with $c = \frac{2}{3}$. The vertex is at $(-5, 0)$, the focus is at $F\left(-5, \frac{2}{3}\right)$, the directrix is $y = -\frac{2}{3}$, and the axis is $x = -5$. The ends of the focal chord are $\left(-5 \pm \frac{4}{3}, \frac{2}{3}\right)$. The graph of the parabola is shown, along with its focus F and directrix.

27. Since the directrix is a vertical line and the focus is to the left of the directrix, the parabola opens to the left. Its equation, then, has the form $(y - k)^2 = 4c(x - h)$, where $c < 0$. The vertex is halfway between the focus and the directrix, so $h = 0$ and $k = 0$. Since the distance from the focus to the vertex is $|c|$, we have $c = -4$. Therefore, the equation of the parabola is

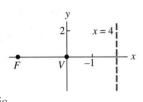

$$(y - 0)^2 = 4(-4)(x - 0) \quad \text{or} \quad y^2 = -16x.$$

30. Since the axis is the line through the focus and the vertex, and the vertex is above the focus, the parabola opens down. Its equation, then, has the form $(x - h)^2 = 4c(y - k)$, where $c < 0$. The vertex is $(0, 0)$ and, since the distance from the vertex to the focus is $|c|$, we have $c = -10$. Therefore, the equation of the parabola is

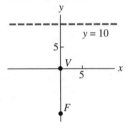

$$(x - 0)^2 = 4(-10)(y - 0) \quad \text{or} \quad x^2 = -40y.$$

33. Since the directrix is a vertical line and the focus is to the left of the directrix, the parabola opens to the left. Its equation, then, has the form $(y - k)^2 = 4c(x - h)$, where $c < 0$. The vertex is halfway between the focus and the directrix, so $h = 2$ and $k = 4$. Since the distance from the focus to the vertex is $|c|$, we have $c = -3$. Therefore, the equation of the parabola is

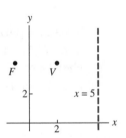

$$(y - 4)^2 = 4(-3)(x - 2) \quad \text{or} \quad (y - 4)^2 = -12(x - 2).$$

36. Since the axis is the line through the focus and the vertex, and the vertex is above the focus, the parabola opens down. Its equation, then, has the form $(x - h)^2 = 4c(y - k)$, where $c < 0$. The vertex is $(-2, 5)$ and, since the distance from the vertex to the focus is $|c|$, we have $c = -2$. Therefore, the equation of the parabola is

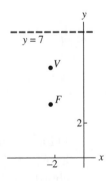

$$(x + 2)^2 = 4(-2)(y - 5) \quad \text{or} \quad (x + 2)^2 = -8(y - 5).$$

39. Since the directrix is a horizontal line and the vertex is above the directrix, the parabola opens up. Its equation, then, has the form $(x - h)^2 = 4c(y - k)$, where $c > 0$. The vertex is halfway between the focus and the directrix, so the focus is $(0, \frac{7}{4})$. Since the distance from the vertex to the directrix is $|c|$, we have $c = \frac{7}{4}$. Therefore, the equation of the parabola is

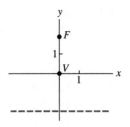

$$(x - 0)^2 = 4\left(\frac{7}{4}\right)(y - 0) \qquad \text{or} \qquad x^2 = 7y.$$

42. Since the directrix is a vertical line and the focus is to the left of the directrix, the parabola opens to the left. Its equation, then, has the form $(y - k)^2 = 4c(x - h)$, where $c < 0$. The vertex is halfway between the focus and the directrix, so the focus is $(-2, 4)$. Since the distance from the focus to the vertex is $|c|$, we have $c = -1$. Therefore, the equation of the parabola is

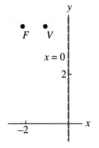

$$(y - 4)^2 = 4(-1)(x + 1) \qquad \text{or} \qquad (y - 4)^2 = -4(x + 1).$$

45. To find the x-intercepts, we set $y = 0$ and solve for x:

$$(0 + 4)^2 = 4(x + 1)$$
$$4 = x + 1$$
$$x = 3.$$

The x-intercept is $(3, 0)$. To find the y-intercepts, we set $x = 0$ and solve for y:

$$(y + 4)^2 = 4(0 + 1) = 4$$
$$y + 4 = \pm 2$$
$$y = \pm 2 - 4.$$

The y-intercepts are $(0, -6)$ and $(0, -2)$.

48. To find the x-intercepts, we set $y = 0$ and solve for x:

$$0 - 8(0) - x + 15 = 0$$
$$x = 15.$$

The x-intercept is $(15, 0)$. To find the y-intercepts, we set $x = 0$ and solve for y:

$$y^2 - 8y - 0 + 15 = 0$$
$$y^2 - 8y + 15 = 0$$
$$(y - 3)(y - 5) = 0.$$

The y-intercepts are $(0, 3)$ and $(0, 5)$.

51. Since light emanating from the focus of a parabola is reflected along a line parallel to the axis of the parabola (the x-axis in this case), the equation of the reflected ray is $y = -2$, where $x \geq 1$.

54. Let the ground be the x-axis and suppose the vertex of the parabola is on the y-axis at $(0, 20)$. Then the equation of the parabola is $x^2 = 4c(y - 20)$, where $c < 0$. Since the point $(4, 18)$ is on the parabola, we have

$$4^2 = 4c(18 - 20)$$
$$4 = c(-2)$$
$$c = -2.$$

Thus, the equation of the parabola is $x^2 = -8(y - 20)$. The water hits the ground at an x-intercept of the parabola. Setting $y = 0$ in the equation of the parabola, we have $x^2 = -8(0 - 20) = 160$, so $x = \pm\sqrt{160} = \pm 4\sqrt{10}$. The water hits the ground $4\sqrt{10} \approx 12.65$ m from the point on the ground directly beneath the end of the pipe.

57. **(a)** For the parabola $x^2 = 8y = 4(2)y$, we identify $c = 2$. Since the vertex is at the origin, the parabola opens up around the y-axis. The ends of the focal chord are at $(-2c, c)$ and $(2c, c)$, so the focal width of the parabola is $2(2) - (-2)(2) = 8$.

(b) The parabolas $x^2 = 4cy$ and $y^2 = 4cx$ have the same size and shape, so their focal widths are the same, and we consider the parabola $x^2 = 4cy$. This is a parabola with vertex at the origin and axis on the y-axis. The focus is at $(0, c)$, so the ends of the focal chord are at $(-2c, c)$ and $(2c, c)$. The focal width of the parabola is $|2c - (-2c)| = 4|c|$.

6.2   The Ellipse

3. We identify $h = 0$, $k = 0$, $a = \sqrt{16} = 4$, and $b = \sqrt{1} = 1$. Since the form of the equation is

$$\frac{x^2}{b^2} + \frac{y^2}{a^2} = 1,$$

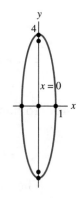

the major axis is vertical. From $b^2 = a^2 - c^2$ we have $c^2 = a^2 - b^2 = 16 - 1 = 15$, so $c = \sqrt{15}$. We use $e = c/a$ to find the eccentricity.

Center: $(0, 0)$;   *Foci:* $(0, -\sqrt{15}), (0, \sqrt{15})$;   *Vertices:* $(0, -4), (0, 4)$;

Endpoints of Minor Axis: $(-1, 0), (1, 0)$;   *Eccentricity:* $e = \dfrac{\sqrt{15}}{4}$.

6. Dividing by 4, we write the equation as

$$\frac{x^2}{2} + \frac{y^2}{4} = 1.$$

We identify $h = 0$, $k = 0$, $a = \sqrt{4} = 2$, and $b = \sqrt{2}$. Since the form of the equation is

$$\frac{x^2}{b^2} + \frac{y^2}{a^2} = 1,$$

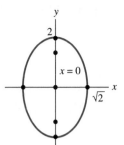

the major axis is vertical. From $b^2 = a^2 - c^2$ we have $c^2 = a^2 - b^2 = 4 - 2 = 2$, so $c = \sqrt{2}$. We use $e = c/a$ to find the eccentricity.

Center: $(0,0)$;   *Foci*: $(0,-\sqrt{2}), (0,\sqrt{2})$;   *Vertices*: $(0,-2), (0,2)$;

Endpoints of Minor Axis: $(-\sqrt{2},0), (\sqrt{2},0)$;   *Eccentricity*: $e = \dfrac{\sqrt{2}}{2}$.

9. We identify $h = 1$, $y = 3$, $a = \sqrt{49} = 7$, and $b = \sqrt{36} = 6$. Since the form of the equation is

$$\frac{(x-h)^2}{a^2} + \frac{(y-k)^2}{b^2} = 1,$$

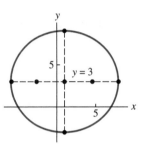

the major axis is horizontal. From $b^2 = a^2 - c^2$ we have $c^2 = b^2 - a^2 = 49 - 36 = 13$, so $c = \sqrt{13}$. We use $e = c/a$ to find the eccentricity.

Center: $(1,3)$;   *Foci*: $(1-\sqrt{13},3), (1+\sqrt{13},3)$;   *Vertices*: $(-6,3), (8,3)$;

Endpoints of Minor Axis: $(1,-3), (1,9)$;   *Eccentricity*: $e = \dfrac{\sqrt{13}}{7}$.

12. We identify $h = 3$, $k = -4$, $a = \sqrt{81} = 9$, and $b = \sqrt{64} = 8$. Since the form of the equation is

$$\frac{(x-h)^2}{b^2} + \frac{(y-k)^2}{a^2} = 1,$$

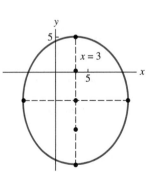

the major axis is vertical. From $b^2 = a^2 - c^2$ we have $c^2 = a^2 - b^2 = 81 - 64 = 17$, so $c = \sqrt{17}$. We use $e = c/a$ to find the eccentricity.

Center: $(3,-4)$;   *Foci*: $(3,-4-\sqrt{17}), (3,-4+\sqrt{17})$;

Vertices: $(3,-13), (3,5)$;   *Endpoints of Minor Axis*: $(-5,-4), (11,-4)$;

Eccentricity: $e = \dfrac{\sqrt{17}}{9}$.

15. Dividing by 45, we write the equation as

$$\frac{(x-1)^2}{9} + \frac{(y+2)^2}{15} = 1.$$

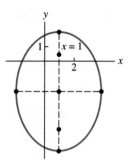

We identify $h = 1$, $k = -2$, $a = \sqrt{15}$, and $b = \sqrt{9} = 3$. Since the form of the equation is

$$\frac{(x-h)^2}{b^2} + \frac{(y-k)^2}{a^2} = 1,$$

the major axis is vertical. From $b^2 = a^2 - c^2$ we have $c^2 = a^2 - b^2 = 15 - 9 = 6$, so $c = \sqrt{6}$. We use $e = c/a$ to find the eccentricity.

Center: $(1, -2)$;   *Foci*: $(1, -2 - \sqrt{6}), (1, -2 + \sqrt{6})$;

Vertices: $(1, -2 - \sqrt{15}), (1, -2 + \sqrt{15})$;

Endpoints of Minor Axis: $(-2, -2), (4, -2)$;   *Eccentricity*: $e = \dfrac{\sqrt{6}}{\sqrt{15}} = \sqrt{\dfrac{2}{5}}$.

18. Factoring 9 from the x^2 and x terms and 5 from the y^2 and y terms, and then completing the square, we have

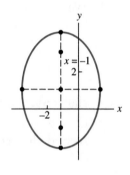

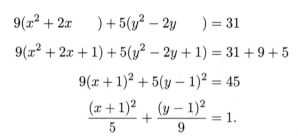

$$9(x^2 + 2x \quad) + 5(y^2 - 2y \quad) = 31$$

$$9(x^2 + 2x + 1) + 5(y^2 - 2y + 1) = 31 + 9 + 5$$

$$9(x+1)^2 + 5(y-1)^2 = 45$$

$$\frac{(x+1)^2}{5} + \frac{(y-1)^2}{9} = 1.$$

We identify $h = -1$, $k = 1$, $a = \sqrt{9} = 3$, and $b = \sqrt{5}$. Since the form of the equation is

$$\frac{(x-h)^2}{b^2} + \frac{(y-k)^2}{a^2} = 1,$$

the major axis is vertical. From $b^2 = a^2 - c^2$ we have $c^2 = a^2 - b^2 = 9 - 5 = 4$, so $c = 2$. We use $e = c/a$ to find the eccentricity.

Center: $(-1, 1)$;   *Foci*: $(-1, -1), (-1, 3)$;   *Vertices*: $(-1, -2), (-1, 4)$;

Endpoints of Minor Axis: $(-1 - \sqrt{5}, 1), (-1 + \sqrt{5}, 1)$;   *Eccentricity*: $e = \dfrac{2}{3}$.

21. Since the vertices are at $(\pm 5, 0)$ and the foci are at $(\pm 3, 0)$, we identify $a = 5$ and $c = 3$. Then $b^2 = a^2 - c^2 = 25 - 9 = 16$ and $b = 4$. The major axis is the x-axis, so the equation of the ellipse is

$$\frac{x^2}{25} + \frac{y^2}{16} = 1.$$

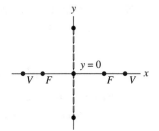

24. Since the vertices are at $(0, \pm 7)$ and the foci are at $(0, \pm 3)$, we identify $a = 7$ and $c = 3$. Then $b^2 = a^2 - c^2 = 49 - 9 = 40$ and $b = 2\sqrt{10}$. The major axis is the y-axis, so the equation of the ellipse is

$$\frac{x^2}{40} + \frac{y^2}{49} = 1.$$

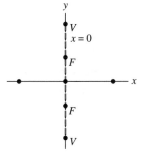

27. Since the center of the ellipse is the midpoint of the major axis, the center is at $(1, -3)$. We then identify $h = 1$, $k = -3$, $a = \frac{1}{2}[5 - (-3)] = 4$, and $b = \frac{1}{2}[-1 - (-5)] = 2$. The equation of the ellipse is

$$\frac{(x-1)^2}{4^2} + \frac{(y+3)^2}{2^2} = 1$$

or

$$\frac{(x-1)^2}{16} + \frac{(y+3)^2}{4} = 1.$$

30. Since the center is at the origin and a focus is at $(1, 0)$, we have $c = 1$. Using $a = 3$ we see that $b^2 = a^2 - c^2 = 3^2 - 1^2 = 8$. The equation of the ellipse is

$$\frac{x^2}{9} + \frac{y^2}{8} = 1.$$

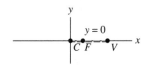

33. The equation is centered at the origin with its major axis along the y-axis, so the form of its equation is

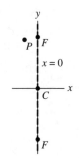

$$\frac{x^2}{b^2} + \frac{y^2}{a^2} = 1 \quad \text{or} \quad \frac{x^2}{a^2 - c^2} + \frac{y^2}{c^2} = 1.$$

Since the distance from the center to a focus is $c = 3$ and the ellipse passes through $P(-1, 2\sqrt{2})$, we have

$$\frac{(-1)^2}{a^2 - 9} + \frac{(2\sqrt{2})^2}{a^2} = 1$$

$$a^2 + 8(a^2 - 9) = a^2(a^2 - 9)$$

$$9a^2 - 72 = a^4 - 9a^2$$

$$a^4 - 18a^2 + 72 = 0$$

$$(a^2 - 6)(a^2 - 12) = 0.$$

Since $a^2 = 6$ implies that $a < 3$, we cannot have $a^2 - 6 = 0$. From $a^2 = 12$ we see that the equation of the ellipse is

$$\frac{x^2}{3} + \frac{y^2}{12} = 1.$$

36. Since the center is at $(1, -1)$ and a focus is at $(1, 1)$, the major axis is vertical and $c = 1 - (-1) = 2$. The form of the equation is

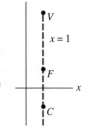

$$\frac{(x - 1)^2}{b^2} + \frac{(y + 1)^2}{a^2} = 1.$$

From $a = 5$ and $c = 2$ we have $b^2 = a^2 - c^2 = 25 - 4 = 21$. Thus, the equation of the ellipse is

$$\frac{(x - 1)^2}{21} + \frac{(y + 1)^2}{25} = 1.$$

39. Since the endpoints of the minor axis are at $E_1(0, -1)$ and $E_2(0, 5)$, the center is at $(0, 2)$ and $b = 5 - 2 = 3$. Because the focus at $(6, 2)$ is $c = 6$ units from the center, we have $a^2 = b^2 + c^2 = 3^2 + 6^2 = 45$. Thus, the equation of the ellipse is

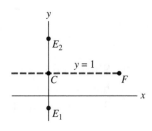

$$\frac{x^2}{a^2} + \frac{(y - 2)^2}{b^2} = 1 \quad \text{or} \quad \frac{x^2}{45} + \frac{(y - 2)^2}{9} = 1.$$

42. From Problem 41, we identify $a = \frac{1}{2}(72) = 36$ million and $b = \frac{1}{2}(70.4) = 35.2$ million. Then

$$c^2 = a^2 - b^2 = 36^2 - 35.2^2 = 1296 - 1239.04 = 56.96,$$

and $c = \sqrt{56.96} \approx 7.5472$. The eccentricity is

$$e = \frac{c}{a} = \frac{7.5472}{36} \approx 0.2096.$$

[*Note:* The eccentricity is actually

$$\frac{7.5472 \text{ million}}{36 \text{ million}} \approx 0.2096,$$

so there are no units associated with the eccentricity.]

45. Referring to the figure, we see that $a = 15$ and $b = 5$, so the equation of the ellipse is

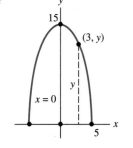

$$\frac{x^2}{5^2} + \frac{y^2}{15^2} = 1 \quad \text{or} \quad \frac{x^2}{25} + \frac{y^2}{225} = 1.$$

Letting $x = 3$ and solving for y, we have

$$\frac{3^2}{25} + \frac{y^2}{225} = 1$$
$$\frac{y^2}{225} = 1 - \frac{9}{25}$$
$$y^2 = 225 - 9(9) = 144.$$

Thus, $y = 12$ feet is the height of the arch where $x = 3$.

48. The distance between the foci is $2c$. Identifying $a = \frac{1}{2}(458) = 229$ and $b = \frac{1}{2}(390) = 195$, we have $c^2 = a^2 - b^2 = 229^2 - 195^2 = 14{,}416$, so $2c = 2\sqrt{14{,}416} \approx 240.13$ m.

51. If (x, y) is a point on the ellipse, then $d_1 + d_2 = 12$ where $d_1 = \sqrt{(x - 0)^2 + (y - 2)^2}$ and $d_2 = \sqrt{(x - 8)^2 + (y - 6)^2}$. That is,

$$\sqrt{x^2 + (y - 2)^2} + \sqrt{(x - 8)^2 + (y - 6)^2} = 12$$
$$\sqrt{(x - 8)^2 + (y - 6)^2} = 12 - \sqrt{x^2 + (y - 2)^2}$$
$$(x - 8)^2 + (y - 6)^2 = 144 - 24\sqrt{x^2 + (y - 2)^2} + x^2 + (y - 2)^2$$
$$x^2 - 16x + 64 + y^2 - 12y + 36 = x^2 + y^2 - 4y + 148 - 24\sqrt{x^2 + (y - 2)^2}$$
$$24\sqrt{x^2 + (y - 2)^2} = 16x + 8y + 48$$

$$3\sqrt{x^2 + (y-2)^2} = 2x + y + 6$$

$$9[x^2 + (y-2)^2] = 4x^2 + 4xy + y^2 + 24x + 18y + 36$$

$$9x^2 + 9y^2 - 36y + 36 = 4x^2 + 4xy + y^2 + 24x + 12y + 36$$

$$5x^2 - 4xy + 8y^2 - 24x - 48y = 0.$$

6.3   The Hyperbola

3. From the equation of the hyperbola, we identify $a = \sqrt{64} = 8$ and $b = \sqrt{9} = 3$. Thus, the vertices are $(0, -8)$ and $(0, 8)$. The center of the hyperbola is the origin. From $c^2 = a^2 + b^2 = 64 + 9 = 73$, we see that $c = \sqrt{73}$ and the foci are at $(0, -\sqrt{73})$ and $(0, \sqrt{73})$. Factoring

$$\frac{y^2}{64} - \frac{x^2}{9} = 0$$

we get

$$\left(\frac{y}{8} - \frac{x}{3}\right)\left(\frac{y}{8} + \frac{x}{3}\right) = 0,$$

so the asymptotes are $y = 8x/3$ and $y = -8x/3$. The eccentricity is $e = c/a = \sqrt{73}/8$.

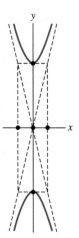

6. Dividing both sides of the equation by 25, we have

$$\frac{x^2}{25} - \frac{y^2}{5} = 1.$$

From this, we identify $a = \sqrt{5}$ and $b = \sqrt{5}$. The center of the hyperbola is the origin, and the vertices are $(-\sqrt{5}, 0)$ and $(\sqrt{5}, 0)$. From $c^2 = a^2 + b^2 = 5 + 5 = 10$, we see that $c = \sqrt{10}$ and the foci are at $(0, -\sqrt{10})$ and $(0, \sqrt{10})$. Factoring

$$\frac{x^2}{5} - \frac{y^2}{5} = 0$$

we get

$$\left(\frac{x}{\sqrt{5}} - \frac{y}{\sqrt{5}}\right)\left(\frac{x}{\sqrt{5}} + \frac{y}{\sqrt{5}}\right) = 0,$$

so the asymptotes are $y = x$ and $y = -x$. The eccentricity is $e = c/a = \sqrt{10}/\sqrt{5} = \sqrt{2}$.

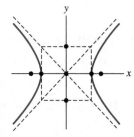

9. From the equation of the hyperbola, we see that the center of the hyperbola is $(5, -1)$. We identify $a = \sqrt{4} = 2$ and $b = \sqrt{49} = 7$. Thus, the vertices are $(3, -1)$ and $(7, -1)$. From $c^2 = a^2 + b^2 = 4 + 49 = 53$, we see that $c = \sqrt{53}$ and the foci are at $(5 - \sqrt{53}, -1)$ and $(5 + \sqrt{53}, -1)$. Factoring

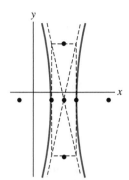

$$\frac{(x-5)^2}{4} - \frac{(y+1)^2}{49} = 0$$

we get

$$\left(\frac{x-5}{2} - \frac{y+1}{7}\right)\left(\frac{x-5}{2} + \frac{y+1}{7}\right) = 0,$$

so the asymptotes are $y = \frac{7}{2}x - \frac{37}{2}$ and $y = -\frac{7}{2}x + \frac{33}{2}$. The eccentricity is $e = c/a = \sqrt{53}/2$.

12. From the equation of the hyperbola, we see that the center of the hyperbola is $(-3, \frac{1}{4})$. We identify $a = \sqrt{4} = 2$ and $b = \sqrt{9} = 3$. Thus, the vertices are $(-3, -\frac{7}{4})$ and $(-3, \frac{9}{4})$. From $c^2 = a^2 + b^2 = 4 + 9 = 13$, we see that $c = \sqrt{13}$ and the foci are at $(-3, \frac{1}{4} - \sqrt{13})$ and $(-3, \frac{1}{4} + \sqrt{13})$. Factoring

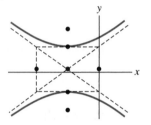

$$\frac{\left(y - \frac{1}{4}\right)^2}{4} - \frac{(x+3)^2}{9} = 0$$

we get

$$\left(\frac{y - \frac{1}{4}}{2} - \frac{x+3}{3}\right)\left(\frac{y - \frac{1}{4}}{2} + \frac{x+3}{3}\right) = 0,$$

so the asymptotes are $y = \frac{2}{3}x + \frac{9}{4}$ and $y = -\frac{2}{3}x - \frac{7}{4}$. The eccentricity is $e = c/a = \sqrt{13}/2$.

15. We first write the equation in the form

$$5(y-7)^2 - 8(x+4)^2 = 40,$$

and then divide both sides by 40. This gives

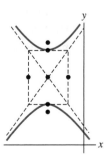

$$\frac{(y-7)^2}{8} - \frac{(x+4)^2}{5} = 1.$$

We see that the center is $(-4, 7)$, and we identify $a = \sqrt{8} = 2\sqrt{2}$ and $b = \sqrt{5}$. Thus, the vertices are $(-4, 7 - 2\sqrt{2})$ and $(-4, 7 + 2\sqrt{2})$.

From $c^2 = a^2 + b^2 = 8 + 5 = 13$, we see that $c = \sqrt{13}$ and the foci are at $(-4, 7 - \sqrt{13})$ and $(-4, 7 + \sqrt{13})$. Factoring

$$\frac{(y-7)^2}{8} - \frac{(x+4)^2}{5} = 0$$

we get

$$\left(\frac{y-7}{\sqrt{8}} - \frac{x+4}{\sqrt{5}}\right)\left(\frac{y-7}{\sqrt{8}} + \frac{x+4}{\sqrt{5}}\right) = 0,$$

so the asymptotes are $y = \sqrt{8/5}\,(x+4) + 7$ and $y = -\sqrt{8/5}\,(x+4) + 7$. The eccentricity is $e = c/a = \sqrt{13/8}$.

18. We first factor 16 from the two x-terms and -25 from the two y-terms:

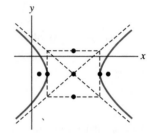

$$16(x^2 - 16x \quad) - 25(y^2 + 6y \quad) = -399.$$

Completing the square in x and y, we get

$$16(x^2 - 16x + 64) - 25(y^2 + 6y + 9)$$
$$= -399 + 1024 - 225 = 400.$$

Dividing both sides by 400, we obtain

$$\frac{(x-8)^2}{25} - \frac{(y+3)^2}{16} = 1.$$

We see that the center is $(8, -3)$, and we identify $a = \sqrt{25} = 5$ and $b = \sqrt{16} = 4$. Thus, the vertices are $(3, -3)$ and $(13, -3)$. From $c^2 = a^2 + b^2 = 25 + 16 = 41$, we see that $c = \sqrt{41}$ and the foci are at $(8 - \sqrt{41}, -3)$ and $(8 + \sqrt{41}, -3)$. Factoring

$$\frac{(x-8)^2}{25} - \frac{(y+3)^2}{16} = 0$$

we get

$$\left(\frac{x-8}{5} - \frac{y+3}{4}\right)\left(\frac{x-8}{5} + \frac{y+3}{4}\right) = 0,$$

so the asymptotes are $y = \frac{4}{5}x - \frac{47}{5}$ and $y = -\frac{4}{5}x + \frac{17}{5}$. The eccentricity is $e = c/a = \sqrt{41}/5$.

21. We identify $c = 5$ and note that $h = 0$ and $k = 0$. Using $a = 5$, we have $a^2 + b^2 = c^2$ or $3^2 + b^2 = 5^2$, so that $b^2 = 25 - 9 = 16$ and $b = 4$. The equation of the hyperbola is

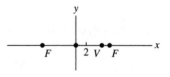

$$\frac{x^2}{9} - \frac{y^2}{16} = 1.$$

24. We identify $c = 3$ and note that $h = 0$ and $k = 0$. Using $a = \frac{3}{2}$, we have $a^2 + b^2 = c^2$ or $(\frac{3}{2})^2 + b^2 = 3^2$, so that $b^2 = 9 - \frac{9}{4} = \frac{27}{4}$ and $b = 3\sqrt{3}/2$. The equation of the hyperbola is

$$\frac{y^2}{9/4} - \frac{x^2}{27/4} = 1.$$

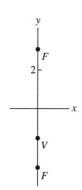

27. We are given $h = 0$, $k = 0$, $a = \frac{5}{2}$, and $c = 3$. Then, using $a^2 + b^2 = c^2$ we have $(\frac{5}{2})^2 + b^2 = 3^2$, so $b^2 = 9 - \frac{25}{4} = \frac{11}{4}$. The equation of the hyperbola is

$$\frac{y^2}{25/4} - \frac{x^2}{11/4} = 1.$$

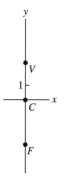

30. We are given $h = 0$, $k = 0$, $a = 1$, and $c = 5$. Then, using $a^2 + b^2 = c^2$ we have $1^2 + b^2 = 5^2$, so $b^2 = 25 - 1 = 24$. The equation of the hyperbola is

$$x^2 - \frac{y^2}{24} = 1.$$

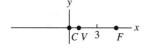

33. The vertices are on the x-axis equidistant from the origin, so the x-axis is the transverse axis and the center of the hyperbola is the origin. Thus, $h = 0$, $k = 0$, and we identify $a = 2$. The form of the equation of the hyperbola is $y = \pm(b/a)x$, so $b/a = 4/3$, $b = \frac{4}{3}a = \frac{4}{3}(2) = \frac{8}{3}$. The equation of the hyperbola is

$$\frac{x^2}{4} - \frac{y^2}{64/9} = 1.$$

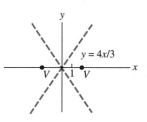

36. We identify $h = 2$, $k = 3$, $c = 2 - 0 = 2$, and $a = 3 - 2 = 1$. The transverse axis is on the horizontal line $y = 3$. Using $a^2 + b^2 = c^2$, we have $1^2 + b^2 = 2^2$, so $b^2 = 4 - 1 = 3$. The equation of the hyperbola is

$$(x - 2)^2 - \frac{(y - 3)^2}{3} = 1.$$

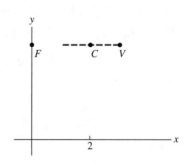

39. The center is at the origin and the transverse axis is on the x-axis, so the form of the equation of the hyperbola is

$$\frac{x^2}{a^2} - \frac{y^2}{b^2} = 1.$$

Since $a = 2$ and $P(2\sqrt{3}, 4)$ is on the parabola,

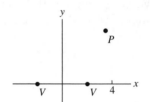

$$\frac{(2\sqrt{3})^2}{2^2} - \frac{4^2}{b^2} = 1$$

$$3 - 1 = \frac{16}{b^2}$$

$$b^2 = \frac{16}{2} = 8.$$

Thus, the equation of the hyperbola is

$$\frac{x^2}{4} - \frac{y^2}{8} = 1.$$

42. We identify $h = 3$, $k = -5$, and $a = -2 - (-5) = 3$. Since the transverse axis is vertical, the equation of the hyperbola has the form

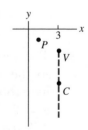

$$\frac{(y + 5)^2}{3^2} - \frac{(x - 3)^2}{b^2} = 1.$$

Since the hyperbola passes through $(1, 1)$,

$$\frac{(-1 + 5)^2}{9} - \frac{(1 - 3)^2}{b^2} = 1$$

$$\frac{16}{9} - 1 = \frac{4}{b^2}$$

$$b^2 = \frac{4}{7/9} = \frac{36}{7}.$$

Thus, the equation of the hyperbola is

$$\frac{(y+5)^2}{9} - \frac{(x-3)^2}{36/7} = 1.$$

45. Let B and C be the foci of a hyperbola passing through G, the location of the gun. We can then identify $h = 0$, $k = 0$, and $c = 2$. The transverse axis of the hyperbola is horizontal, so the form of the equation of the hyperbola is $x^2/a^2 - y^2/b^2 = 1$. Since the gun is 2 km closer to B than to C, we have $2a = 2$, so $a = 1$. Using $a^2 + b^2 = c^2$ we then have $1^2 + b^2 = 2^2$, so $b^2 = 3$. The equation of the hyperbola is $x^2 - y^2/3 = 1$. The gun also lies on the line through $A(-10, 16)$ and $C(2, 0)$. The slope of this line is $m = 16/(-10-2) = -\frac{4}{3}$, and its equation is $y = -\frac{4}{3}(x-2)$. Substituting this into the equation of the hyperbola, we have

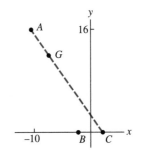

$$x^2 - \frac{[(-4/3)(x-2)]^2}{3} = 1$$
$$3x^2 - \frac{16}{9}(x-2)^2 = 3$$
$$27x^2 - 16(x^2 - 4x + 4) = 27$$
$$11x^2 + 64x - 91 = 0$$
$$(11x - 13)(x + 7) = 0.$$

From the figure, we see that x must be negative, so $x = -7$. This gives $y = -\frac{4}{3}(-7-2) = 12$. The gun is located at $(-7, 12)$.

48. (a) Identifying $a^2 = 4$ and $b^2 = 9$, we have $c^2 = a^2 + b^2 = 4 + 9 = 13$, so $c = \sqrt{13}$. The point on the hyperbola corresponding to $x = \sqrt{13}$ is

$$y = \sqrt{\frac{9}{4}(\sqrt{13})^2 - 9} = 3\sqrt{\frac{13}{4} - 1} = \frac{3}{2}\sqrt{9} = \frac{9}{2}.$$

Thus, the focal width is $2(\frac{9}{2}) = 9$.

(b) Using $c^2 = a^2 + b^2$, the point on the hyperbola corresponding to $x = c$ is

$$y = \sqrt{\frac{b^2}{a^2}c^2 - b^2} = \frac{b}{a}\sqrt{c^2 - a^2} = \frac{b^2}{a}.$$

Thus, the focal width is $2b^2/a$.

6.4   Polar Coordinates

3. We measure $\frac{1}{2}$ unit along the ray $\pi/2 + \pi = 3\pi/2$.

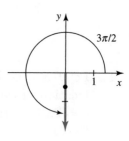

6. We measure $\frac{2}{3}$ unit along the ray $7\pi/4$.

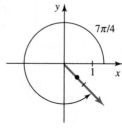

9. **(a)** We use $\theta = -2\pi + \pi/3 = -5\pi/3$, so the point is $(4, -5\pi/3)$.

   (b) We use $\theta = 2\pi + \pi/3 = 7\pi/3$, so the point is $(4, 7\pi/3)$.

   (c) We use $\theta = \pi + \pi/3 = 4\pi/3$, so the point is $(-4, 4\pi/3)$.

   (d) We use $\theta = \pi/3 - \pi = -2\pi/3$, so the point is $(-4, -2\pi/3)$.

12. **(a)** We use $\theta = -2\pi + 7\pi/6 = -5\pi/6$, so the point is $(3, -5\pi/6)$.

   (b) We use $\theta = 2\pi + 7\pi/6 = 19\pi/6$, so the point is $(3, 19\pi/6)$.

   (c) We use $\theta = 7\pi/6 - \pi = \pi/6$, so the point is $(-3, \pi/6)$.

   (d) We want to express the ray given by $\pi/6$ in part (c) as a negative angle. We use $\theta = -2\pi + \pi/6 = -11\pi/6$, so the point is $(-3, -11\pi/6)$.

15. With $r = -6$ and $\theta = -\pi/3$, we have, from (1) in the text,

$$x = -6\cos\left(-\frac{\pi}{3}\right) = -6\left(\frac{1}{2}\right) = -3$$

$$y = -6\sin\left(-\frac{\pi}{3}\right) = -6\left(-\frac{\sqrt{3}}{2}\right) = 3\sqrt{3}.$$

Thus, $(-6, -\pi/3)$ is equivalent to $(-3, 3\sqrt{3})$ in rectangular coordinates.

18. With $r = -5$ and $\theta = \pi/2$, we have, from (1) in the text,

$$x = -5\cos\frac{\pi}{2} = -5(0) = 0$$

$$y = -5\sin\frac{\pi}{2} = -5(1) = -5.$$

Thus, $(-5, \pi/2)$ is equivalent to $(0, 5)$ in rectangular coordinates.

21. With $x = 1$ and $y = -\sqrt{3}$, we have, from (2) in the text,

$$r^2 = 1^2 + (-\sqrt{3})^2 = 4 \qquad \text{and} \qquad \tan\theta = -\sqrt{3}.$$

   (a) For $r > 0$ we take $r = 2$ and for $-\pi < \theta < \pi$ we take $\theta = -\pi/3$. Thus, $(1, -\sqrt{3})$ is equivalent to $(2, -\pi/3)$ in polar coordinates.

   (b) For $r < 0$ we take $r = -2$. Then we use $\theta = -\pi/3 + \pi = 2\pi/3$. Thus, $(1, -\sqrt{3})$ is also equivalent to $(-2, 2\pi/3)$ in polar coordinates.

24. With $x = 1$ and $y = 2$, we have, from (2) in the text,

$$r^2 = 1^2 + 2^2 = 5 \qquad \text{and} \qquad \tan\theta = \frac{2}{1} = 2.$$

   (a) For $r > 0$ we take $r = \sqrt{5}$. Then we use $\theta = \tan^{-1} 2$, which is a first quadrant angle. Thus, $(1, 2)$ is equivalent to $(\sqrt{5}, \tan^{-1} 2)$ in polar coordinates.

   (b) For $r < 0$ we take $r = -\sqrt{5}$ and use the third quadrant angle $\pi + \tan^{-1} 2$ for θ. In this case, $(1, 2)$ is equivalent to $(-\sqrt{5}, \pi + \tan^{-1} 2)$ in polar coordinates.

27. Substituting $x = r\cos\theta$ and $y = r\sin\theta$ into the given equation, we find

$$r\sin\theta = 7r\cos\theta$$
$$r(\sin\theta - 7\cos\theta) = 0.$$

Since $r = 0$ is just the pole, the polar form of the line $y = 7x$ is

$$\sin\theta - 7\cos\theta = 0$$
$$\sin\theta = 7\cos\theta$$
$$\frac{\sin\theta}{\cos\theta} = 7$$
$$\tan\theta = 7,$$

or $\theta = \tan^{-1} 7$.

30. Substituting $x = r\cos\theta$ and $y = r\sin\theta$ into the given equation, we find

$$r^2\cos^2\theta - 12r\sin\theta - 36 = 0$$
$$r^2(1 - \sin^2\theta) = 36 + 12r\sin\theta$$
$$r^2 = r^2\sin^2\theta + 12r\sin\theta + 36$$
$$r^2 = (r\sin\theta + 6)^2$$
$$r = \pm(r\sin\theta + 6).$$

Solving the two equations for r gives

$$r = \frac{6}{1 - \sin\theta} \qquad \text{or} \qquad r = \frac{-6}{1 + \sin\theta}. \qquad (6.1)$$

Recall that $\sin(\theta + \pi) = -\sin\theta$. If $(-r, r+\pi)$ is substituted into the second equation, we find

$$-r = \frac{-6}{1 - \sin\theta} \qquad \text{or} \qquad r = \frac{6}{1 - \sin\theta}.$$

Thus, the two equations in (6.1) are equivalent and we take the polar equation to be $r = 6/(1 - \sin\theta)$.

33. We use $r^2 = x^2 + y^2$ and $x = r\cos\theta$. Substituting into the given equation, we have

$$r^2 + r\cos\theta = r \qquad \text{or} \qquad r(r + \cos\theta - 1) = 0,$$

where $r > 0$ since $\sqrt{x^2 + y^2}$ is positive. Then $r = 0$ and $r = 1 - \cos\theta$. When $\theta = 0$, $\cos\theta = 1$ and $r = 0$, so we need only use $r = 1 - \cos\theta$ as the polar version of the rectangular equation $x^2 + y^2 + x + \sqrt{x^2 + y^2}$.

6.5   Graphs of Polar Equations

3. The graph of $\theta = \pi/3$ is a line through the origin that makes an angle of $\pi/3$ with the positive x-axis.

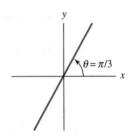

6. The graph of $r = 3\theta$, $\theta \geq 0$, is a spiral.

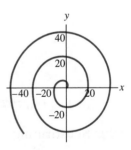

9. The graph of $r = 2(1 + \sin\theta)$ is a cardioid.

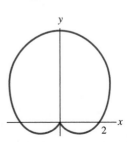

12. Identifying $a = 2$ and $b = 4$ in $r = 2 + 4\sin\theta$, we have $a/b = \frac{1}{2} < 1$, so the graph is a limaçon with an interior loop.

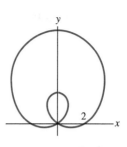

15. Identifying $a = 4$ and $b = 1$ in $r = 4 + \cos\theta$, we have $a/b = 4 > 2$, so the graph is a convex limaçon.

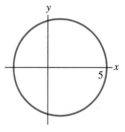

18. The graph of $r = 3\sin 4\theta$ is a rose curve. Since 4 is an even number, there are $2(4) = 8$ petals.

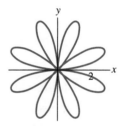

21. The graph of $r = \cos 5\theta$ is a rose curve. Since 5 is an odd number, there are 5 petals.

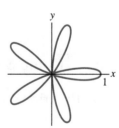

24. The graph of $r = -2\cos\theta$ is a circle centered at $(-1, 0)$.

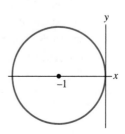

27. The graph of $r^2 = 4\sin 2\theta$ is a lemniscate.

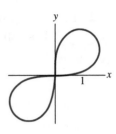

30. Since r^2 is nonnegative, $\sin 2\theta \leq 0$, so $\pi < 2\theta < 2\pi$ and $\pi/2 < \theta < \pi$. In this case, $r = \pm 3\sqrt{-\sin 2\theta}$.

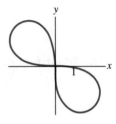

33. The two equations imply $4\sin\theta = 2$, so $\sin\theta = \frac{1}{2}$ and $\theta = \pi/6$, $5\pi/6$. The points of intersection are $(2, \pi/6)$ and $(2, 5\pi/6)$.

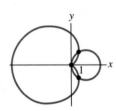

36. Adding the two equations, we obtain $2r = 3$, so $r = \frac{3}{2}$. This implies $\cos\theta = \frac{1}{2}$ or $\theta = \pi/3$, $5\pi/3$. The points of intersection are $(3/2, \pi/3)$ and $(3/2, 5\pi/3)$. From the figure, we see that $(0, 0)$ is also a point of intersection.

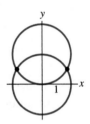

6.6   Conic Sections in Polar Coordinates

3. By writing the equation as

$$r = \frac{4}{1 + \frac{1}{4}\cos\theta},$$

we see that the eccentricity is $e = \frac{1}{4}$, and so the conic is an ellipse. Since the equation has the form given in (5) in the text, the axis of the ellipse lies along the x-axis. In polar form, the vertices are where $\theta = 0$ and $\theta = \pi$, so they are at $(2, 0)$ and $(\frac{16}{3}, \pi)$ in polar coordinates. The y-intercepts occur where $\theta = \pi/2$ and $\theta = 3\pi/2$, so they are at $(4, \pi/2)$ and $(4, 3\pi/2)$.

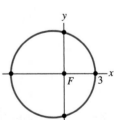

6. By writing the equation as

$$r = \frac{4}{1 - \cos\theta},$$

we see that the eccentricity is $e = 1$, and so the conic is a parabola. Since the equation has the form given in (5) in the text, the axis of the parabola lies along the x-axis. In polar form, the vertex is where $\theta = \pi$, so it is at $(2, \pi)$ in polar coordinates. The y-intercepts occur where $\theta = \pi/2$ and $\theta = 3\pi/2$, so they are at $(4, \pi/2)$ and $(4, 3\pi/2)$.

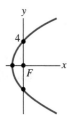

9. From

$$r = \frac{6}{1 - \cos\theta},$$

we see that the eccentricity is $e = 1$, and so the conic is a parabola. Since the equation has the form given in (5) in the text, the axis of the parabola lies along the x-axis. In polar form, the vertex is where $\theta = \pi$, so it is at $(3, \pi)$ in polar coordinates. The y-intercepts occur where $\theta = \pi/2$ and $\theta = 3\pi/2$, so they are at $(6, \pi/2)$ and $(6, 3\pi/2)$.

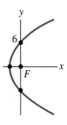

12. By writing the equation as

$$r = \frac{5}{1 - \frac{3}{2}\cos\theta},$$

we see that the eccentricity is $e = \frac{3}{2}$, and so the conic is a hyperbola. Now, write the equation in the form

$$r - \frac{3}{2}r\cos\theta = 5 \qquad \text{or} \qquad r = 5 + \frac{3}{2}r\cos\theta.$$

Substituting $r = \pm\sqrt{x^2 + y^2}$ and $x = r\cos\theta$ and simplifying, we have

$$\pm\sqrt{x^2 + y^2} = 5 + \frac{3}{2}x$$

$$x^2 + y^2 = 25 + 15x + \frac{9}{4}x^2$$

$$y^2 - \frac{5}{4}x^2 - 15x = 25$$

$$y^2 - \frac{5}{4}(x^2 + 12x \qquad = 25$$

$$y^2 - \frac{5}{4}(x^2 + 12x + 36) = 25 - \frac{5}{4}(36) = -20$$

$$\frac{(x+6)^2}{16} - \frac{y^2}{20} = 1.$$

Identifying $a^2 = 16$ and $b^2 = 20$, we have $c^2 = a^2 + b^2 = 36$. Thus, $c = 6$ and $e = c/a = \frac{6}{4} = \frac{3}{2}$.

15. Since the directrix is a vertical line to the right of the origin, the equation of the conic has the form $r = ep/(1+e\cos\theta)$. Since $e = 1$, the conic is a parabola, and since the directrix is 3 units from the origin, $p = 3$. Thus, the equation of the parabola is

$$r = \frac{1(3)}{1+1\cos\theta} = \frac{3}{1+\cos\theta}.$$

18. Since the directrix is a vertical line to the right of the origin, the equation of the conic has the form $r = ep/(1+e\cos\theta)$. Since $e = \frac{1}{2}$, the conic is an ellipse, and since the directrix is 4 units from the origin, $p = 4$. Thus, the equation of the ellipse is

$$r = \frac{\frac{1}{2}(4)}{1+\frac{1}{2}\cos\theta} = \frac{4}{2+\cos\theta}.$$

21. Because the conic is a parabola, the eccentricity is $e = 1$. Since the vertex lies on the y-axis below the origin, the equation of the parabola has the form $r = p/(1-\sin\theta)$. The distance from the focus to the directrix is p and the vertex is halfway between the two, so $p = 2(\frac{3}{2}) = 3$ and

$$r = \frac{3}{1-\sin\theta}.$$

24. Because the conic is a parabola, the eccentricity is $e = 1$. Since the vertex lies on the x-axis to the right of the origin, the equation of the parabola has the form $r = p/(1+\cos\theta)$. The distance from the focus to the directrix is p and the vertex is halfway between the two, so $p = 2(2) = 4$ and

$$r = \frac{4}{1+\cos\theta}.$$

27. We are given $r_a = 12{,}000$ and $e = 0.2$. We let r_p be the perigee distance and solve equation (7) in the text for r_p:

$$0.2 = \frac{r_a - r_p}{r_a + r_p}$$
$$0.2r_a + 0.2r_p = r_a - r_p$$
$$1.2r_p = r_a - 0.2r_a$$
$$r_p = \frac{0.8r_a}{1.2} = \frac{0.8(12{,}000)}{1.2} \approx 8{,}000 \text{ km}.$$

30. **(a)** We are given $2a = 3.34 \times 10^9$, so $a = 1.67 \times 10^9$. Also, we are given $e = 0.97$, so $c = ae = (1.67 \times 10^9)(0.97) \approx 1.62 \times 10^9$. Now, the distance from the focus at $(0,0)$ to the vertex on the negative x-axis is $a - c$. Thus,

$$r(\pi) = \frac{0.97p}{1-0.97(-1)} = \frac{0.97p}{1.97} = a - c = 1.67 \times 10^9 - 1.62 \times 10^9 = 0.05 \times 10^9,$$

so

$$0.97p = 1.97 \times 0.05 \times 10^9 = 0.0985 \times 10^9$$

and

$$p = \frac{0.0985 \times 10^9}{0.97} \approx 0.1 \times 10^9 = 10^8.$$

The equation of the orbit is

$$r = \frac{0.97 \times 10^8}{1 - 0.97 \cos\theta}.$$

(b) $r_a = r(0) = \dfrac{0.97 \times 10^8}{1 - 0.97} = \dfrac{0.97 \times 10^8}{0.03} \approx 32.3 \times 10^8 = 3.23 \times 10^9$ mi

$r_p = r(\pi) = 0.05 \times 10^9 = 5 \times 10^7$ mi

6.7   Calculus Preview—Parametric Equations

3.

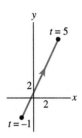

6.

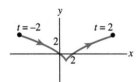

9.

12. Substituting $t^3 + t = x - 4$ into y, we obtain

$$y = -2(x - 4) = -2x + 8.$$

15. From $y = 3\ln t$ we have $t = e^{y/3}$. Substituting this into x, we obtain

$$x = t^3 = (e^{y/3})^3 = e^y \qquad \text{or} \qquad y = \ln x,$$

where $x > 0$.

18. Squaring both sides of $x + 1 = \cos t$ and $y - 2 = \sin t$, we obtain $(x+1)^2 = \cos^2 t$ and $(y-2)^2 = \sin^2 t$. Adding the equations, we get

$$(x+1)^2 + (y-2)^2 = 1.$$

21.

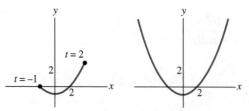

We see that the parameterized curve is only a portion of the rectangular graph.

24.

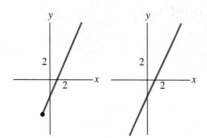

We see that the parameterized curve is a ray lying on the line $y = 2x - 2$.

27.

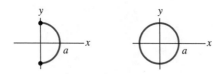

$$x = a \cos t \qquad\qquad x = a \cos 2t$$
$$y = a \sin t \qquad\qquad y = a \sin 2t$$

30. We find the x-intercepts by solving

$$y = t^2 + t - 6 = 0 \qquad \text{or} \qquad y = (t+3)(t-2) = 0.$$

Thus, the x-intercepts occur for $t = -3$ and $t = 2$. When $t = -3$, $x = 6$, and when $t = 2$, $x = 6$. The only x-intercept is $(6, 0)$. We find the y-intercepts by solving

$$x = t^2 + t = 0 \qquad \text{or} \qquad x = t(t+1) = 0.$$

Thus, the y-intercepts occur for $t = 0$ and $t = -1$. When $t = 0$, $y = -6$, and when $t = -1$, $y = -6$. The only y-intercept is $(0, -6)$.

33. Setting $v_0 = 190$, $\theta_0 = 45°$, and $g = 32$, by (1) in the text, parametric equations of the ball's path are

$$x = (190 \cos 45°)t \approx 134.35t$$

and

$$y = -\frac{1}{2}(32)t^2 + (190\sin 45°)t \approx -16t^2 + 134.35t, \quad t \geq 0.$$

When $t = 2$, $x = 268.7$ and $y = -64 + 268.7 = 204.7$, so the ball is at the point $(268.7, 204.7)$.

36. Refer to the figure at the right. Using triangle BCO, we have $\cot\theta = x/2a$ or $x = 2a\cot\theta$. Since angle OAC subtends an arc of $180°$, OAC is a right triangle and $\sin\theta = h/2a$ or $h = 2a\sin\theta$. Now, using triangle ODA, $\sin\theta = y/h$. Thus, $y = h\sin\theta = 2a\sin^2\theta$. The parametric equations are $x = 2a\cot\theta$ and $y = 2a\sin^2\theta$.

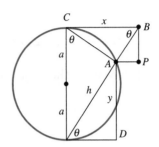

Chapter 6 Review Exercises

3. Since the directrix is a horizontal line and the vertex is halfway between the focus and the directrix, the vertex is at $(1, -5)$. The focus is above the directrix, so the parabola opens up. Because c is the distance from the vertex to the focus, $c = 2$. Thus, the equation of the parabola is $(x - 1)^2 = 8(y + 5)$.

6. We write the equation of the parabola in the form $(x+4)^2 = \frac{1}{8}(y-2) = 4(\frac{1}{32})(y-2)$.

    The vertex of the parabola is $(-4, 2)$. The parabola opens up and $c = \frac{1}{32}$, so the focus is $(-4, \frac{65}{32})$.

9. The transverse axis of the hyperbola is vertical and the center is $(-3, 0)$. Since $a = 1$, the vertices are $(-3, -1)$ and $(-3, 1)$.

12. We identify $a^2 = 1$ and $b^2 = 1$, so $c^2 = a^2 + b^2 = 2$ and $c = \sqrt{2}$. The eccentricity is $e = c/a = \sqrt{2}/1 = \sqrt{2}$.

15. Identifying $x = 0$ and $y = -10$, we see that the point is 10 units down the negative y-axis. Thus, polar coordinates of the point are $(10, 3\pi/2)$.

18. Solving $x = t + 2$ for t, we obtain $t = x - 2$. Substituting into $y = 3 + \frac{1}{2}t$, we get $y = 3 + \frac{1}{2}(x - 2) = \frac{1}{2}x + 2$. This is a line, and since $-\infty < t < \infty$, the parametric equations represent the entire line.

21. Polar coordinates are not unique, but rectangular coordinates are unique, so the statement is true.

24. The center of the hyperbola is (h, k) and the asymptotes of a hyperbola always pass through the center of the hyperbola, so the statement is true.

27. The asymptotes of $x^2 - y^2/25 = 1$ are obtained from

$$x^2 - \frac{y^2}{25} = \left(x - \frac{y}{5}\right)\left(x + \frac{y}{5}\right) = 0,$$

while the asymptotes of

$$\frac{y^2}{25} - x^2 = \left(\frac{y}{5} - x\right)\left(\frac{y}{5} + x\right) = 0.$$

Since the asymptotes are the same, the statement is true.

30. Writing the equation in the form

$$r = \frac{90}{15 - \sin\theta} = \frac{6}{1 - \frac{1}{15}\sin\theta},$$

we see that the eccentricity is $e = \frac{1}{15}$. Since the eccentricity is close to 0, the graph of the ellipse is nearly circular and the statement is true.

33. For $r = 2 + 4\sin\theta$, we identify $a = 2$ and $b = 4$. Since $a/b = \frac{1}{2} < 1$, the graph is a limaçon with an interior loop and the statement is true.

36. Writing $x - 1 = \cos t$ and $y - 1 = \sin t$, we see that $(x-1)^2 + (y-1)^2 = \cos^2 t + \sin^2 t = 1$, so the graph is a circle centered at $(1, 1)$ having radius 1. The statement is true.

39. Since $r^2 = x^2 + y^2$ and $y = r\sin\theta$, a polar equation having the same graph as $x^2 + y^2 - 4y = 0$ is $r^2 - 4r\sin\theta = 0$ or $r(r - 4\sin\theta) = 0$. From $r = 4\sin\theta$, we see that when $\theta = 0$, $r = 0$, so we do not need to use $r = 0$ from $r(r - 4\sin\theta) = 0$. The polar equation is $r = 4\sin\theta$.

42. Since the vertices lie on the y-axis, the form of the equation in polar coordinates is $r = ep/(1 \pm e\sin\theta)$. We are given $e = 2$ and both vertices are below the origin, so $r = 2p/(1 - 2\sin\theta)$, and we need only find p. From $r(\pi/2) = -2p$ and $r(3\pi/2) = 2p/3$ we have $-2p = -4$ so $p = 2$, and $2p/3 = 4/3$, so $p = 2$. The equation of the hyperbola is $r = 4/(1 - 2\sin\theta)$.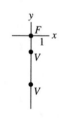

45. Dividing both sides of the equation by 16, we have

$$\frac{(x-1)^2}{16} + \frac{(y-1)^2}{4} = 1.$$

This is an ellipse whose major axis is horizontal, with $a = 4$ and $b = 2$. The center is $(1, 1)$ and the vertices are $(1 \pm 4, 1)$ or $(-3, 1)$ and $(5, 1)$. The endpoints of the minor axis are $(1, 1 \pm 2)$ or $(1, -1)$ and $(1, 3)$. To find the foci we use $a^2 = b^2 + c^2$, so $c^2 = a^2 - b^2 = 16 - 4 = 12$ and $c = 2\sqrt{3}$. The foci are $(1 \pm 2\sqrt{3}, 1)$ or $(1 - 2\sqrt{3}, 1)$ and $(1 + 2\sqrt{3}, 1)$.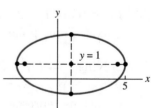

48. We rewrite the equation by moving the constant term and x-term to the right-hand side of the equation and then complete the square on the y-terms:

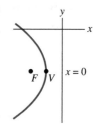

$$y^2 + 10y + = -8x - 41$$
$$y^2 + 10y + 25 = -8x - 41 + 25 = -8x - 16$$
$$(y + 5)^2 = 4(-2)(x + 2).$$

This is the equation of a parabola opening to the left with vertex $(-2, -5)$. Since $c = -2$, the focus is 2 units to the left of the vertex at $(-4, -5)$. The directrix is the y-axis or $x = 0$.

51. The graph is a circle centered at the origin with radius 5.

54. The graph is a circle with radius $\frac{2}{2} = 1$ and center on the x-axis at $(-1, 0)$.

57. This is the equation of a limaçon with $a = 2$ and $b = 1$, so that $a/b = 2 > 1$ and the curve is a convex limaçon.

60. This is the polar form of a conic with eccentricity $e = 1$, so it is a parabola opening to the left.

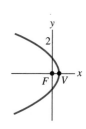

63. Refer to Figure 6.6.7 in the text. We are given that $a = 1 \times 10^9$ and $b = 3 \times 10^8$. Since $a^2 = b^2 + c^2$ for an ellipse,

$$c^2 = a^2 - b^2 = (1 \times 10^9)^2 - (3 \times 10^8)^2 = 1 \times 10^{18} - 9 \times 10^{16}$$
$$= 100 \times 10^{16} - 9 \times 10^{16} = 91 \times 10^{16} = 9.1 \times 10^{17}$$

and $c = \sqrt{9.1 \times 10^{17}} \approx 9.54 \times 10^8$. The distance from the satellite to the center of the planet is

$$a + c = 1 \times 10^9 + 9.54 \times 10^8 = 10 \times 10^8 + 9.54 \times 10^8$$
$$= 19.54 \times 10^8 = 1.954 \times 10^9 \text{ m}.$$

PART V

Final Examination Answers

1. $2\left(x+\frac{3}{2}\right)^2 + \frac{1}{2}$

2. 12

3. $(-5,-3] \cup [0,3] \cup (5,\infty)$

4. $-a+3$

5. ± 16

6. fourth

7. **(a)** $(1,-7)$; **(b)** $(-1,7)$; **(c)** $(-1,-7)$

8. -27; 3

9. $f(x) = x\left(x-1-\sqrt{7}\right)\left(x-1+\sqrt{7}\right)$

10. -1 and -2

11. $\pi/4$

12. -4π

13. $\ln 48$

14. $x = -\frac{5}{2}$

15. false

16. false

17. false

18. false

19. true

20. false

21. true

22. true

23. false

24. true

25. false

26. false

27. true

28. false

29. true

30. true

31. true

32. false

33. (i) and (a);   (ii) and (c);
    (iii) and (d);   (iv) and (b)

34. $(-\infty,-2) \cup \left(-\frac{8}{3},\infty\right)$

35. yes

36. second and fourth

37. (c)

38. (i) and (d);   (ii) and (c);
    (iii) and (b);   (iv) and (a)

39. the interval $(0,10)$

40. $(-2,0) \cup (0,\infty)$

41. $y = -\frac{5}{8}x$

42. $f(x) = -10x^2 + 20x - 6$

43. $\dfrac{4}{\sqrt{x^6+4}+x^3}$

44. $f(x) = 5x^{8/3} - 4x^{13/6} + 8x^{-1/3}$

45. $f(x) = \dfrac{13}{x} + \dfrac{6}{x^2} - \dfrac{6}{x-1}$

46. $f(x) = \sec^2 x - \sec x \tan x$

47. $f(x) = x^3$

48. $f(x) = \begin{cases} x^2 - 3x, & x \le 0 \text{ and } x \ge 3 \\ -x^2 + 3x, & 0 < x < 3 \end{cases}$

49. $x = 0,\ x = -\sqrt{\frac{8}{3}},\ x = \sqrt{\frac{8}{3}}$

50. $x = -\pi,\ x = -\pi/3,\ x = 0,\ x = \pi,$
$x = \pi/3$

51. $\dfrac{15}{(2x + 2h + 5)(2x + 5)}$

52. $-3x^2 - 3xh - h^2 + 20x + 10h$

53. $8;\ [0, 6]$

54. $-1/\sqrt{6}$

55. 0

56. $x = e^{n\pi},\ n$ an integer

57. $(3, 0),\ (0, -6),\ (0, -2)$

58. $(-1, 5);\ (-2, 5),\ (0, 5);$
$(-1 - \sqrt{2}, 5),\ (-1 + \sqrt{2}, 5);$
$(-1, 4),\ (-1, 6)$

59. $(-1, 3)$

60. approximately $46.6°$

61. 1.36 g

62. $0,\ \pi/4,\ \pi/2,\ 3\pi/4,\ \pi,\ 5\pi/4,\ 3\pi/2,$
$7\pi/4,\ 2\pi$